MW00365389

HOW MEDICINE WORKS AND WHEN IT DOESN'T

HOW MEDICINE WORKS AND WHEN IT DOESN'T

Learning Who to Trust to Get and Stay Healthy

F. Perry Wilson, MD

GRAND CENTRAL

NEW YORK BOSTON

Grand Central Publishing
Hachette Book Group
1290 Avenue of the Americas, New York, NY 10104
grandcentralpublishing.com
twitter.com/grandcentralpub

First Edition: January 2023

Grand Central Publishing is a division of Hachette Book Group, Inc. The Grand Central Publishing name and logo is a trademark of Hachette Book Group, Inc.

Library of Congress Cataloging-in-Publication Data
Names: Wilson, F. Perry, author.
Title: How medicine works and when it doesn't : learning who to trust to get and stay healthy / F. Perry Wilson, MD MSCE.
Description: First edition. | New York : Grand Central Publishing, 2023. | Includes bibliographical references.
Identifiers: LCCN 2022037037 | ISBN 9781538723609 (hardcover) | ISBN 9781538723623 (ebook)
Subjects: LCSH: Physician and patient—United States—Popular works. | Medical care—United States—Popular works.
Classification: LCC R727.3 .W553 2023 | DDC 610.69/6—dc23/eng/20220816
LC record available at https://lccn.loc.gov/2022037037

ISBNs: 9781538723609 (hardcover), 9781538723623 (ebook)

Printed in the United States of America

LSC-C

Printing 1, 2022

To Buck and Pam Wilson, for setting the example

Contents

CONTENTS

Introduction

I LOST MS. MEYER twenty-five minutes into her first visit.

Doctors are often a bit trepidatious meeting a patient for the first time. By the time we open the door to the exam room, we've read through your chart, looked at your blood work, and made some mental notes of issues we want to address. Some of the more sophisticated practices even have a picture of you in the electronic medical record, so we have a sense of what you look like. I usually take a beat before I open the door, a quick moment to forget my research lab, my paperwork, a conversation with a coworker, to turn my focus to you, the patient, waiting in that room. It is my hope, standing just on the other side of an inch of wood, that you and I will form a bond, or, more aptly, a "therapeutic alliance." I've always liked that term—the idea that you and I are on the same side of some great war, that together we can overcome obstacles. But that alliance doesn't come easily. And lately, it has been harder to forge than ever.

Ms. Meyer was standing in the center of the room, arms crossed. Smartly dressed and thin, she lived in one of the affluent Philadelphia suburbs—on "the Main Line"—and it showed, in her subtle but clearly expensive jewelry as well as her demeanor. She looked out of place in my resident-run medical clinic, which primarily catered to less wealthy inhabitants of West Philadelphia. But what struck me most was the emotion that radiated from her. Ms. Meyer was angry.

"What brought you here today?" I asked her, using my standard first question. Later in my career, I would learn to replace that line with something more open: "How can I help you?" or even "Tell me about yourself." But it hardly mattered.

She was exhausted, she said. Almost no energy. So drained she could barely get out of bed. Unable to focus during the day, she tossed and turned all night and repeated the cycle day in and day out. It was, she said, simply untenable. I asked how long it had been happening.

"Months," she said. "Years, actually. You are literally the sixth doctor I've seen about this." Her anger broke to reveal desperation.

Second opinions are common enough in medical practice. Third opinions, for difficult cases, are not unheard of. But I had never been a sixth opinion before, and I felt immediately uncomfortable. Not because I wasn't confident in my diagnostic abilities—like all young doctors I hadn't yet learned how much I didn't know—but because I was worried that whatever thoughts I had about her possible ailment would not be enough. What could I offer that all these others couldn't?

I kept my poker face firmly intact and waited.

Eleven seconds. That's how long the typical doctor waits before interrupting a patient, according to a study in the *Journal of General Internal Medicine*. Determined to *not* be a typical doctor, I let her talk, in her own words and in her own time. I thought my attentive listening would frame our relationship differently—that she might see me as a physician who was conscientious, methodical. But it backfired. It was clear she resented the fact that she had to relay the same information to me that she had already told to the five doctors that came before me.

One of the most important skills a doctor has is to read the room. So I switched from respectful listening to diagnosing. I tried to troubleshoot symptoms of possible thyroid dysfunction, anemia, sleep apnea, lymphoma and other cancers. I asked about her family history, her history of drug or alcohol abuse, her sexual history. I even made sure I didn't miss questions pertaining to pregnancy, because (this one comes from experience) you should never assume someone *isn't* pregnant. I reviewed her lab work: Pages upon pages of blood and urine tests. Even CT scans of the head, chest, abdomen, and pelvis. Nothing was out of order. Nothing that we can measure in a lab or in the belly of a CT scanner, at least.

But her affect was off, and her mood was sad. Ms. Meyer seemed, frankly, depressed. There is a formal way to diagnose major depressive disorder; a patient must display five of nine classic symptoms (such as loss of interest in activities they used to enjoy, fatigue, or weight changes). Ms. Meyer had eight of nine, a clear-cut case of major depression, according to the diagnostic manuals. But was it depression? Or was it something else, and the frustration of living with that something else had led to depression?

The nine classic symptoms are far from the only way depression can manifest. As a disease that lives in the brain, the symptoms can be legion—and can lead doctors and patients on costly, and often fruitless, wild-goose chases.

"Listen," I said, "not everything is super-clear-cut in Medicine. I think part of this might be a manifestation of depression. It's really common. Maybe we should try treating that and seeing if your energy improves."

Right there. That's when I lost her.

I could tell from the set of her jaw, the way her eyes stopped looking directly at mine and flickered off a bit, centering on my forehead. I could tell from her silence, and from the slight droop in her posture, that she had lost hope. We talked some more, but the visit was over. There would be no therapeutic alliance. I asked her to call the number on the back of her insurance card to set up a consultation with a mental health professional and made her a follow-up appointment with me in a month, which she, unsurprisingly, missed. My rush to a diagnosis—in this case a diagnosis that comes with a stigma (unwarranted, but a stigma nonetheless)—drove her away from both me and from conventional medicine. And had she even heard a diagnosis at all? Or had she heard, like so many women have about so many concerns over so many years, "It's all in your head"?

I didn't see her for another year. When I did, she was having a seizure in the emergency room, the result of a "water cleanse" a naturopathic practitioner had prescribed. Forcing herself to drink gallons of water a day, she had diluted the sodium content in her blood. When her sodium level got too low, her brain could not appropriately send electrical signals, and seizures ensued. She would survive, thankfully, and tell me later that she had never felt better. She had been told all her problems were due to heavy metal toxicity. (Lab work would not confirm this.) This diagnosis had led her into a slew of questionable medical practices, including regular "cleanses" and chelation therapy—where substances similar to what you might find in water softening tablets are injected into the blood to bind harmful metals. Chelation therapy runs around $10,000 to $20,000 per year and is not covered by insurance.

The striking thing was that she positively shone with confidence and hope. Lying in a hospital bed, recovering from life-threatening seizures, she was, in a word, happy.

And I felt... Well, to be honest, I think the emotion I felt was jealousy. It would be one thing if no one could help poor Ms. Meyer, depressed and unwilling to even entertain the diagnosis, but someone *did* help her. Someone whose worldview was, in my mind, irrational at best and exploitative at worst. My instinct was to dismiss Ms. Meyer as another victim of an industry of hucksters, as a rube. She had been taken in with empty promises and false hope, and some grifter had extracted cash from her in the manner of televangelists and late-night psychic hotlines. His "treatment" landed her in the emergency room with generalized tonic-clonic seizures that could have killed her. This was bad medicine, plain and simple.

But—and this "but" was why I continue to think about Ms. Meyer—in the way that mattered to *her*, she got better. The huckster helped.

It took me a long time to figure out why—fifteen years, actually. In that time, I finished my residency and fellowship at the University of Pennsylvania. I got a master's degree in clinical epidemiology (the study of how diseases affect a population). I was brought onto the faculty at Yale University and started a research lab running clinical trials to try and generate the hard data that would really save lives. I became a scientist and a researcher, and a physician caring for the sickest of the sick. I lectured around the world on topics ranging from acute kidney injury to artificial intelligence and published more than one hundred peer-reviewed medical manuscripts. And yet, somehow, I knew that all the research studies I did would be for nothing if I couldn't figure out how I—how Medicine—had failed Ms. Meyer and all the people out there who feel abandoned, ignored by the system, or overwhelmed by medical information.

Why were people turning to their family and friends or social media for medical advice when physicians are willing and able to provide the best possible information? Was it simply the cost of healthcare? Or was something deeper going on? And though it took time, what I figured out will shine a light on why doctors have lost touch with their patients, why patients have lost faith in their doctors, and how we can get back to that therapeutic alliance that we all need in order to be truly healthy. That is what this book is all about.

It turns out the most powerful force in Medicine is not an antibiotic. It isn't stem cell therapy, genetic engineering, or robotic surgery. The most powerful force in Medicine is trust. It is the trust that lives between a patient and a physician, and it goes both ways. I trust you to tell me the truth about how you feel and what you want. You trust me to give you the best advice I can possibly give. We trust each other to fight against whatever ails you, physical or mental, to the best of our abilities. Ms. Meyer did not trust me. That was *my* failure, not hers. And that personal failure is a mirror of the failure of Medicine writ large—our failure to connect with patients, to empathize, to believe that their ailment is real and profound, and to honestly explain how medical science works and succeeds, and why it sometimes doesn't. We doctors have failed to create an environment of trust. And into that vacuum, others have stepped.

It's not entirely doctors' fault, of course. The average primary care physician has less than fifteen minutes to conduct a typical new-patient visit. If the doctor doesn't stick to that time, the practice will go out of business—overwhelmed by payments for malpractice insurance, overhead, and dwindling reimbursements from insurers. It's hard to create trust in fifteen minutes. Combine our limited schedules with a seemingly unfeeling healthcare system, which sometimes charges thousands of dollars for an ambulance ride to the hospital and tens of thousands of dollars for even routine care, and it is no wonder why, according to a study in the *New England Journal of Medicine*, trust in physicians is lower in the United States than in twenty-three other economically developed countries.

While the healthcare system and physicians are not synonymous, physicians are the face of that system. In earlier times, we ran that system. It is no longer the case. Most physicians haven't realized this yet, but we are no longer a managerial class. We are labor, plain and simple, working for others who, without medical training but with significant business acumen, use our labor to generate profit for companies and shareholders. Part of the key to restoring trust between patients and doctors is for all of us to start fighting to reform the system. And doctors should be on the front line of that battle.

There is a right way and a wrong way to earn someone's trust. One key lesson in this book is that it takes a keen observer to tell the difference. Honesty, integrity, transparency, validation: These are good ways to create trust, and physicians need to commit to them wholeheartedly if we ever want our patients to take us seriously. Patients need to commit to honesty and transparency as well, even when the truth is painful. But less-than-scrupulous individuals can also leverage certain cognitive biases to create trust in ways that are manipulative. Trust hacking like this is a central reason modern medicine has lost ground to others who promise a quick fix for what ails you. It's important not only to evaluate your own methods, but also to be able to spot whether someone is trying to earn your trust in an ethical way, to spot bad actors whose intentions may have little to do with actually helping you.

There are several ways to hack trust. One is to give an impression of certainty. The naturopath who treated Ms. Meyer was unambivalent. He told her *exactly* what was wrong with her: heavy metal toxicity. There was no long list of potential alternative diagnoses, no acknowledgment of symptoms that were typical or atypical for that diagnosis. He provided clarity and, through that, an impression of competence. To know who you can *truly* trust, you have to learn to recognize this particular trick—you have to be skeptical of people who are overly certain, overly confident. Health is never clear-cut; nothing is 100 percent safe and nothing is 100 percent effective. Anyone who tells you otherwise is selling something. This book will show you how to grapple with medical uncertainty and make rational decisions in the face of risk.

Traditional doctors like me are trained early on to hedge their bets. Patients *hate* this. Ask a doctor if the medication you are being prescribed will work, and they will say something like "For most people, this is quite effective" or "I think there's a good chance" or (my personal pet peeve) "I don't have a crystal ball." This doctorly ambivalence is born out of long experience. We all have patients who do well, and we all have patients who do badly. We don't want to lie to you. We're doing the best we can. And, look, I know that this is frustrating.

Neil deGrasse Tyson, the astronomer and brilliant science communicator, once wrote, "The good thing about Science is that it's true, whether or not you believe in it." When it comes to the speed of light, the formation of nebulae, and the behavior of atoms, this is true. The laws of the universe are the laws of the universe; they "change" only insofar as our tools to study them have improved. But Medicine is not astrophysics. It is not an exact science. Or if it is, we have not yet explored enough of the nooks and crannies of the human machine to be able to fix it perfectly.

Physicians, if we are being honest, will admit that their best advice is still a guess. A very *good* guess—informed by years of training and centuries of trial and error. But we are still playing the odds. Trust hackers, though, are never so equivocal. Ask your local homeopath how to cure your headaches, and you will be told they have just the thing.

You can also hack trust by telling people what they want to hear. For someone who is sick, tell them they will be cured. For someone who is dying, tell them they will live. For someone who feels a stigma surrounding their depression, tell them it is not their own brain, but an external toxin, that is wreaking havoc. To know who to trust with your health, you need to first know yourself. You need to know, deep down, what you *want* to be true. And be careful of those who tell you it *is* true.

This skill, consciously avoiding the cognitive bias known as "motivated reasoning" (the tendency to interpret facts in a way that conforms with your desired outcome), is challenging for all of us—doctors included. But it is probably the most critical skill to have if you want to make the best, most rational choices about your health. The answer you are looking for might *not* be the right answer. That's why we will discuss, right in the first chapter, how before you know who else to trust, you have to learn to trust yourself.

The community of people vying for your trust is truly massive. It spans individuals from your neighbors and your friends on social media to the talking heads on the nightly news. All of them are competing in a trust marketplace, and not all of them are playing fair. A smattering of recent headlines illustrates the overwhelming amount of medical-sounding "facts" you may have been exposed to: COFFEE CURES CANCER; DEPRESSED MOTHERS GIVE BIRTH TO AUTISTIC CHILDREN; MARIJUANA IS A GATEWAY TO OPIATE ABUSE; EGGS INCREASE THE RISK OF HEART DISEASE; EGGS DECREASE THE RISK OF HEART DISEASE. Each day, we are inundated with confusing and conflicting headlines like these, designed more to shock, sell, and generate clicks than to inform. I will give you the skills to figure out what health information can be trusted and what is best left unliked and unretweeted.

The information age brought with it the promise of democratization of truth, where knowledge could be accessed and disseminated at virtually no cost by anyone in the world. But that promise has been broken. Instead, the information age has taught us that data is cheap but *good* data is priceless. We are awash in bad data, false inference, and "alternative facts." In that environment, we are all—doctors and patients alike—subject to our deepest biases. We are able to look for "facts" that fit the narrative of our lives, and never forced to question our own belief systems. If we can't interrogate the quality of the information we're consuming, we can't make the best choices about our health. It's that simple.

When you read this book, you'll learn that doctors aren't perfect. As humans, we have our own biases. Rigorous studies have shown that those biases lead to differential treatment by race, sexual orientation, and body mass index. While most physicians are worthy of your trust, not all of them are. I'll teach you how to recognize those who aren't putting your interests first.

It's not wrong to be skeptical of Medicine. Medical science has been developing, evolving, and advancing for the past one hundred years, and has had many stumbles along the way. Scandals from the repressing of information about harms linked to Vioxx (a drug that was supposed to relieve pain), to the effects of thalidomide in pregnancy (which was designed to reduce nausea but led to severe birth defects), to the devastating heart problems caused by the diet pill fen-phen remind us that the profit motive can corrupt the best science. Alleged frauds like the linking of the measles, mumps, and rubella (MMR) vaccine to autism diagnoses pollute the waters of inquiry, launch billion-dollar businesses, and leave the public unsure of what to really believe.

Why would I, a physician and researcher, highlight the failures of medical research? Because Medicine isn't perfect or complete. It is also, in terms of the alleviation of human suffering, the single greatest achievement of humankind. But you need to understand Medicine, warts and all, to make the right choices about your own health. We must be skeptical, but never cynical.

This book will also detail some of the astounding successes and breakthroughs that medical science has made possible. For the vast majority of human history, life-or-death issues were determined by randomness or chance. Maybe it was a broken bone that prevented someone from hunting and gathering, or a cut on the arm that got infected, or a childbirth that developed complications for the mother and her child. It's no mystery why before the modern era, one in four babies died before their first birthday. And those who survived their first year had only a fifty-fifty chance of reaching adulthood. These days, the script has been flipped. Ninety-five percent of humans born on Earth today will reach adulthood, and life expectancy has more than doubled in the last two hundred years. We've witnessed the near eradication of diseases like smallpox, rubella, and polio, which would have easily killed or disabled our ancestors, and we've achieved major advances in drug treatment and medical procedures that can prolong our lives despite the onset of deadly diseases. Medical science, translated from lab bench to bedside to the doctor's prescription pad, has been nothing short of miraculous. It has transformed the human experience from lives that are, to steal from Thomas Hobbes, "nasty, brutish and short," to the lives we live today, which, while not without their troubles, would be unrecognizable to our ancestors.

Here we stand, in the midst of a torrent of information that would have been inconceivable thirty years ago. Some of it is good, some is bad, but all is colored by our own biases and preconceptions. Decisions about your health happen every single day. If you want to be in control, you need to know how to separate the good from the bad, whether it comes from someone sitting atop the ivory tower, or from your friend on Facebook. This book is about medical science. But it's really about learning to trust again. When you finish reading it, you will no longer be swayed by the loudest voice, the most impassioned plea, or the most retweeted article. You will be able to trust your doctor, trust yourself, and trust Medicine—our imperfect science and the single greatest force for good in the world today.

Some Notes on What Follows

I wrote this book to help people understand the complexity, the beauty, the challenges, and the failures of modern medicine. As you read, you'll come across multiple stories of patients I have encountered in my medical practice. Unless I had direct permission from the patient, I changed certain details to make it impossible to identify them in order to respect their privacy. This includes factors like their name, their age, and details of their medical care that I considered unique enough to allow reidentification by those familiar with that care. None of these changes alter the fundamental details of what happened to them, or to me, as we interacted.

Why so many stories? One of the problems with understanding Medicine in the modern era is an overreliance on anecdote in favor of hard data. This is exacerbated by a social media apparatus that amplifies anecdotes far more efficiently than dry old data sets. Anecdotes should never form the basis of a medical decision. But anecdotes are not without value. Human beings are storytellers. Stories are the most efficient way to convey the cultural and social values we espouse and share. The stories in this book are meant to frame a discussion, not provide evidence as to the appropriateness of one treatment over another.

Finally, a note on the distinctive treatment of an important word. I have chosen to capitalize "Medicine" when I am referring to the practice (or art, or science) of caring for the health and well-being of other humans. When talking about pills and injections that go into humans, I use lowercased "medicine." The difference between these two concepts deserves more than an alternate capitalization scheme, but it is not my place to upend the current lexicon.

Thank you for joining me. Let's get started.

HOW MEDICINE WORKS AND WHEN IT DOESN'T

CHAPTER 1

Our Most Human Failing

I MET GARLIC MAN two weeks into my first year of medical school.

The first two years of medical school are spent in classrooms and libraries, poring over books and memorizing lists of body parts, medications, and physiologic pathways—the biochemical equivalent of "The leg bone is connected to the foot bone." In contrast to college, med school is an intimate affair, with around one hundred students per year and no real electives. For classes, we sat in the same lecture hall all day; the professors were the ones who rotated through. Two straight years in one chair, more or less, surrounded by the same people. It's a wonder more of us don't drop out.

Years three and four are "clinical," when students venture into hospitals and clinics and begin to interact with real patients. But Columbia University's College of Physicians and Surgeons, my alma mater, had started to mix it up a bit. Believing that an early taste of real clinical work would keep us engaged, motivated, and enthusiastic, it sent the first-year class off to do various patient-adjacent activities in the areas around New York City's Washington Heights neighborhood.

I was assigned to the local senior center, which occupied the basement floor of a six-story building on Fort Washington Avenue, a mostly residential street, broken up by the occasional bodega or arepa restaurant, and a bagel shop I still miss because of its amazing olive cream cheese. My job at the center was to take blood pressures. The key qualification was that I owned my own stethoscope and had passed a training session that lasted fifteen minutes.

My orientation to the center was equally brief. I was shown a table and a folding chair at the front of a long common room and told to sit. The room itself was filled with folding chairs and circular plastic tables (also foldable), giving me the impression that at a moment's notice the place could be cleared out for a square dance or a junior prom. While I was there, though, most of the residents were sitting at the tables, playing cards, chatting, reading, or simply being around others. The manager told me to make myself comfortable and left. I would not see him again for the next six months. The senior center residents knew what to do, though. I was, apparently, only the latest in a long line of bright-eyed med students who had been given this job. As soon as I sat down, short white coat nicely pressed and stethoscope around my neck like the doctors on TV, people would start shuffling into a line about forty-five deep.

It was a lot of blood pressures to take in the hour I was there. And, given that the majority of my time with a patient was spent with my stethoscope in my ears, there wasn't much time for chitchat. I would say hello, wrap the cuff around their arm, and inflate, noting the Korotkoff sounds that would inform me what the systolic and diastolic pressures were. I'd write them down on a slip of paper—systolic over diastolic—and give them back to the patient. Without any real medical training, I couldn't answer questions, so if the numbers seemed too high, I would ask them to talk to their doctor and not forget to take their medications as prescribed.

Week after week, I began to recognize faces. Without names to attach to them, I created little nicknames in my head. There was

Debonair Greek Guy, with the slicked-back hair. There was Old Frail Lady, whose arm was so thin the blood-pressure cuff kept sliding off. And there was Garlic Man.

Garlic Man (née Sid), age eighty-six, was so denominated in my mind not only because of the scent that permeated his clothing and person, but because of his evangelism. He preached the gospel of garlic and he lived as he preached—taking fifteen hundred milligrams in tablet form twice a day and cooking with "as much garlic as possible." And I, twenty-two years old and surrounded by octogenarians, was more than a receptive audience. Garlic Man was charismatic, charming, and direct. Where the other residents shuffled into line, he strode. Where they stumbled or slurred their speech, he enunciated. Week after week, he extolled the virtues of the miracle plant. He took no other medications, had no medical problems. His blood pressure was, of course, perfect. Garlic was the answer to every health concern.

And I was looking for an answer. Medical school introduces young people to the concept of mortality through the anatomy lab—where teams of five students are assigned a preserved cadaver, which they slowly dissect over the course of six months. It can be a traumatic experience; one of my classmates fainted as soon as the shroud was lifted from her cadaver. Another would faint almost half a year later, when we started to dissect the hands. This was common, I was told. People faint when they see the cadaver's face and when they see its hands, two parts that remind us that the bodies are truly human. So mortality was on my mind when Garlic Man told me that, essentially, death is a choice we make by not eating enough garlic.

I won't say I was overly credulous, but I was intrigued enough to do some research. Papers out of China suggested that garlic consumption was associated with a reduced risk of cancer. American papers linked eating garlic with improved cholesterol and a reduced risk of heart attack. What's more, I *liked* garlic. Increasing my garlic intake would be a pleasure, not a chore. And I did, perfecting a garlic bread recipe I use to this day and crushing clove after clove into soups, pastas, and sauces. I began to be something of a Garlic Man myself, which my then girlfriend (now wife)... tolerated.

Week after week, Garlic Man and I would talk together, almost always about the pungent plant, and (though doctors shouldn't play favorites) he quickly became the patient I most looked forward to seeing. And then one week, he wasn't there. This shouldn't have worried me—blood-pressure checks were optional, but he had never missed one before. And two weeks later, I saw him again, near the end of the line of my forty-five patients. But he looked different: disheveled, a bit haggard, diminished. As he came closer, not making eye contact, I noticed a urine stain on the front of his trousers.

"Are you okay?" I asked as he fell into the chair on the other side of the table. The smell of garlic was there, but so was something else—a sourness that suggested he hadn't showered in a few days.

His wife had Alzheimer's disease, he told me, and she no longer recognized him. She wasn't eating much, and the doctors felt that her lack of appetite and worsening mental condition meant that this was the end.

I didn't know what to say. Twenty years and thousands of patients later, I still don't know what to say. There were people behind him in line. We were in a public space. So I did what I always do: I took his blood pressure and wrote down the numbers. As he rose to leave, I said, "I'm sorry about your wife, Sid."

"Yeah, me too."

He walked away, head still bowed, and it hurt me that I couldn't help him.

You learn early in medical school that the word "doctor" comes from the Latin *docere*—to teach. We are supposed to be teachers, helping our patients learn how best to care for themselves. But I was not a teacher for Garlic Man. Quite the opposite. He had been teaching me. Sure, he had taught me about garlic and the multiple ways you can cram more of it into your diet. But in that moment, seeing him powerless, frail, and scared, he taught me something much more important: That we are not what we want to be. We are not what others need us to be. We are simply what we are. We are mortal. We lose those we love. We age. We die. The battle against death is not lost only by the cadavers in the anatomy lab, but by every human that has ever lived. And no amount of garlic will change that fact.

It hurt to see Sid, my role model for graceful aging, decline in the wake of his wife's worsening condition. But at the same time, it forced me to reflect on what had been so compelling about him. Sid presented a version of reality that I wanted to believe was true. That there was a simple solution to my fear of my own mortality. That I could hold it at bay with a few extra squeezes of the garlic press. It was that desire that prevented me from initially seeing the deeper truth: Here was a man who had been suffering silently, papering over the pain of slowly losing his wife by keeping a stiff upper lip and lecturing a captive med student about garlic. It was a constructed fantasy, one that both of us were eager to believe.

When we want something to be true, we are more likely to believe it is true. What's more, we will interpret the facts we are given in such a way that they support that underlying belief. This is the key premise of the concept of motivated reasoning, and once you are aware of it, you will see it everywhere. It explains why certain people believe that climate change is not human-made, or that President Barack Obama wasn't really born in the United States. It explains why we play the lottery, why we eat "just one more" potato chip, and why we don't believe weather forecasts that call for rain on our wedding day. It explains why some smokers will tell you that nicotine isn't addictive or that smoking doesn't cause lung cancer. It affects doctors and patients alike and limits our ability to make truly informed choices.

It sounds simple, I know, but recognizing what it is you want to be true can be devilishly hard. Although this book promises to show you who to trust with your medical and health decisions, we start here, inside our own minds, because the first step to knowing who to trust when you are making medical decisions is to be able to trust yourself.

The Conclusion You Want, the Facts That Fit

"Facts" are definitive statements. They can be true (*It is sunny outside*) or false (*I can run a five-minute mile*). We know this intuitively, and so we use multiple techniques to determine the veracity of facts. Our brains have what amounts to an internal fact-checking system, which uses a variety of data, both conscious and unconscious, to assign a truth value to a fact or concept. We believe facts that our brains determine are likely to be true and disbelieve those that our brains determine are likely to be false.

But the system is imperfect. It has limited access to information. It can be hacked. When it is, we believe things that are not true. Knowing how the system can be hacked can help us make it stronger and more readily reject the falsehoods that come at us at breakneck speed in the modern world.

We all like to consider ourselves rational people. When presented with a fact, we tell ourselves we evaluate it in the context of everything else we know to be true, look at supporting and contravening evidence, and come to an objective, informed conclusion. But our brains are not the fact-checking department at National Public Radio, looking for citations and calling in experts to provide their opinions. Rather, our interpretation of facts tends to be based not on the reliability of the fact itself, but on the desirability of the conclusion that leads from the fact. Our brains are fact sculptors, molding and shaping the facts we are given to fit an underlying narrative that *feels* true, or a conclusion that we believe *should* be true. Facts that match a narrative we already believe in seem more reliable by that quality alone. This process of motivated reasoning is not just pervasive; it is universal—it is built into the way human beings think. It can lead us to bad decisions about our health, and, more than that, it can lead us down some pretty dark rabbit holes.

Many ideas that we could call conspiracy theories are really just extreme examples of motivated reasoning—taking facts and interpreting them to get to a conclusion that one wants to be true. When I discuss this with friends and colleagues, I am often met with some befuddlement. "You're saying people *want* the US government to have secretly planned to destroy the World Trade Center towers?"

Actually, yes. The presence of a superpowerful controlling force, even if evil, at the very least imposes order on a world that may otherwise seem chaotic. That a ragtag bunch of terrorists could cause so much death and destruction in the United States implies a certain amount of randomness in the world that would make anyone feel unsafe. But if the destruction was part of a large-scale *plan* by those in power, we can paradoxically rest somewhat easier: The randomness is removed. The world has more structure. Some people need that.

I am no more immune to this than the next guy, though few would describe me as conspiratorial. For example, one of my best friends is a vegan, mostly for ethical reasons. But even putting ethics aside, the facts are pretty clear: People who eat a diet rich in fruits, vegetables, nuts, and legumes tend to have fewer heart attacks, are less likely to develop cancer, and overall live longer. If I were being perfectly rational, I would look at those facts and conclude that it makes sense to not eat meat anymore. But I don't *like* that conclusion. It is not what I *want* to be true. So I tell myself that these dietary studies have flaws (they do—see chapter 3) and therefore I can safely ignore them. But deep down I know that I have a bias here: a conclusion I want to reach, which is that I can continue to eat all the delicious (if unethical and unhealthy) things I always eat.

The fact that I am aware of that bias helps a bit—it allows me to review data about nutrition with a more balanced eye. But I'd be lying if I said the bias isn't there. The way I justify my meat eating and the way a smoker justifies his pack-a-day habit aren't that different. We may know that the facts are against us, but we spin the facts, we interpret them, we decide they don't apply to us for some reason. These are all variants of motivated reasoning, which allows us to come to the conclusion we all want: that we don't have to change.

Motivated reasoning largely defends the status quo. And in Medicine, that can be a real problem.

Motivated Reasoning in Medicine

One of my favorite quantitative examples of motivated reasoning, perhaps because it mimics a typical medical scenario, comes from Kent State University in 1992. In the tradition of many highly cited psychology studies, it recruited psychology majors for course credit and put them through the proverbial wringer.

The students were told that a new saliva-based test had been developed to detect a potentially serious medical condition called "thioamine acetylase deficiency," which, if present, could lead to pancreatic disease in the future. Unbeknownst to the students, TAA deficiency was a fake condition—all part of the psychological manipulation.

The students were given a yellow strip of paper to lick. Half were told it would turn green if they had adequate TAA levels. Half were told it would turn green if they had inadequate TAA levels. (In reality, it was just a yellow piece of paper; it wouldn't be turning green for anyone.)

The students were sent to the bathroom to test themselves and told that if the paper were to change color, it would typically happen in ten seconds. The setup was now complete. Half the students, seeing no color change, would believe they had a potentially dangerous medical condition, while half, also seeing no color change, would be relieved to find out they were fine.

Knowing how motivated reasoning works, you can probably guess how the students reacted. All were motivated to reach the conclusion that they were healthy. (That's something we all wish to be true.) Half got data supporting that belief, while half did not. And the reactions were dramatically different. Those who believed that no color change meant they had TAA deficiency waited much longer before turning their slip in (an average of almost 105 seconds versus almost 76 seconds), presumably hoping the color change was delayed. More dramatically, more than 50 percent of the students who thought no color change was bad retested themselves. Less than 20 percent of students who were led to believe it was good retested.

After the test, the students were asked to rate how serious they believed TAA deficiency is. Those led to believe they had TAA deficiency rated it a 31.7 on a 100-point scale, compared to a 49.8 among those led to believe they didn't have the condition. Again, motivated reasoning. Those who believed they had a new diagnosis of TAA deficiency rationalized that it wasn't so bad.

All students were told the truth before they left the experiment, to much relief. But what happened in that lab happens all the time in medical offices around the country. Our desire to reach the conclusion that we are healthy makes us doubt diagnoses that we don't want to hear, and seek out multiple opinions in the hope that we will find some data that leads us to the conclusion we wish to be true—that all is well, or at least that all is easily fixable.

Just because a particular fact relates to the human body or physiology or a medication doesn't give it any special privilege. There are true medical facts (*The kidney is the only place in the body where a capillary connects two arteries*) and false medical facts (*Peptic ulcers are caused by stress*). Recognizing true facts from false facts is a critical skill, and much of this book will help you separate the wheat from the chaff on that front. But we're not there yet. Because even true facts are subject to that amorphous cognitive process known as "interpretation." And interpretation, if you let it, will be colored by motivated reasoning.

Let me provide you with a true fact, courtesy of the *New England Journal of Medicine*: The risk of homicide is significantly higher in homes with guns. How you interpret that fact may depend on a host of factors: the other studies you've read or heard about, your personal experiences, the last homicide case you saw on the news. But more than any other factor, how you interpret that fact is likely to be influenced by what you think of guns and gun ownership. If you believe that the right to bear arms is a bulwark of human dignity in an uncertain world, then you may interpret the link between gun ownership and homicide as a result of the fact that individuals at higher risk of violence are more likely to buy guns. If you feel that unrestricted gun purchasing poses a public health threat and encourages more violence, you are likely to interpret the link between gun ownership and homicide as strong evidence that guns *cause* people to hurt others. Same fact, different interpretation. That is motivated reasoning.

Motivated reasoning about medical topics is nothing new, but in the summer of 2021, the death toll associated with it began to rise significantly.

Motivated Reasoning and Vaccine Hesitancy

The coronavirus pandemic had raged throughout 2020, keeping my hospital full and my colleagues and me in a near-constant panic around our own safety and that of our families. But by December 2020, the first vaccines received emergency use authorization by the FDA. I likened it, in a live CNN segment, to J. R. R. Tolkien's Battle of Helm's Deep—Gandalf and the Rohirrim arriving just at the moment all hope was lost. (The producers gently encouraged me to make my references a bit less obscure in the future.) Nevertheless, hope was high for the new year, and as the pace of vaccinations increased, it seemed that victory against the virus was within our grasp. Vaccinations rose to five hundred thousand per day, one million per day, two million per day. And then . . . they started to drop.

You know the story. After about half the US population was vaccinated, finding eager arms for vaccine shots became more and more difficult. The remaining unvaccinated individuals were labeled "vaccine-hesitant." Some of my colleagues felt this was too gentle a term and preferred the more aggressive "anti-vax," but in talking with my patients, friends, and even some people who reached out to me online, it was clear that "hesitant" was the right word.

While there were some diehards who would never agree to vaccination and were actively trying to convince others not to get vaccinated, the majority of the hesitant fell into a few camps. There were those who were concerned about vaccines in general—they had a sense that the bar for injecting something, anything, into your body should be a particularly high one. There were some who didn't have an explicit reason to avoid vaccination—they just never seemed to get around to it. And there were some who were particularly skeptical of *these* vaccines, who pointed out that they had been developed and tested in under a year, which represented a compression of the typical five- to ten-year timetable.

I should note here that this is one of those "facts" that isn't exactly true. It was in 2003 that mRNA vaccines hit their stride, after the initial SARS epidemic, when scientists recognized that the spillover of coronaviruses from animal populations into humans could be an ongoing problem. We would have been in much worse shape against COVID-19 if we hadn't had that seventeen-year head start.

There are myriad reasons for vaccine hesitancy, but by far the most common refrain I heard in the summer of 2021 was that we didn't know the long-term risks of the vaccine. This is, empirically, true. We don't know the long-term risks of any new invention, by definition. Honestly, we don't know the long-term risks of COVID-19 either (though they aren't looking good, as the docs at our "long COVID" clinic will tell you). However, we did have some facts about side effects of the vaccine since those early reactions were carefully tracked. Some data came from the rigorous randomized trials of the vaccines, which suggested that mild side effects were more common with a vaccine compared to a placebo, and that serious side effects were quite rare. But the facts most commonly misinterpreted in the era of vaccine hesitancy didn't come from the vaccine trials—they came from the Vaccine Adverse Event Reporting System. VAERS has been around since 1990 and was designed as a mechanism for surveillance of rare vaccine side effects.

Vaccines enjoy a somewhat unique place in Medicine, as one of the few classes of medication that are given to people who are, objectively, not sick. That makes the risk-benefit calculus for vaccines substantially different than it is for other medications. Lingering side effects or complications are simply not acceptable, especially when certain vaccines are recommended for practically everyone.

Healthcare providers are required, by law, to report to VAERS certain events that occur within certain time periods (they vary depending on the vaccine), but anyone can make a VAERS report. This has resulted in some humorous entries, including one from a man who reported that a vaccine had turned him into the Incredible Hulk. Let me give you a piece of data: As of June 13, 2022, there were 871,953 VAERS reports regarding COVID-19 vaccines. There were 120,503 VAERS reports for all other vaccines combined over the same time period. How you interpret these facts will depend strongly on how you feel about vaccines in general, and the COVID vaccine in particular. If you are vaccine-hesitant, you might view these numbers as a startling confirmation of those deep-seated worries you've had all along. Others might look at those numbers and conclude that in a politically charged environment where all we talk about all day is COVID-19 vaccines, it's little wonder that people would report events—whether linked to the vaccine or not.

In fact, the political overtones of COVID vaccination may be responsible both for the poor quality of VAERS data and for our failure to get enough people vaccinated to get the virus under control. Medicine is no stranger to politics (see the debates on universal healthcare and abortion access), but it is actually rare to see discrete healthcare interventions be linked to one political stance or another. You are not more likely to take your cholesterol medicine if you are a Republican or a Democrat. But the political divide with regard to the COVID-19 vaccine is stark: In August 2021, a Kaiser Family Foundation survey found that 51 percent of unvaccinated individuals were Republican, while 23 percent were Democrats. Of individuals who said they had no intention of *ever* getting the vaccine, 58 percent were Republican and 15 percent were Democrats.

There are several reasons we may have sorted this way. It could be due to President Donald Trump's downplaying of the severity of the virus itself, or it could be due to Republicans' more hard-line stance on autonomy. But regardless, there are communities in the United States where admitting you are vaccinated would shock and upset those around you, and communities where the exact opposite is true. Throw an "anyone can report" system like VAERS into that mix, and you get hundreds of thousands of reports.

Many epidemiologists would argue that, for this reason, VAERS data should be treated as essentially useless. But there are no doubt some valuable needles in the VAERS haystack. As it turns out, there are some standardized statistical approaches that can be used to try to tease out some truth from the messy VAERS data. They often involve comparing the rates of reported side effects from one vaccine to another while applying statistical weights to account for the amount of noise in the data. As of this writing, these approaches have suggested that some COVID-19 vaccines carry a small risk (about one in ten thousand) of heart inflammation, and others a similar risk of blood clot formation. Compared to the risk of death from COVID-19 if you are unvaccinated, which is probably around eight or nine in one thousand—the risks from vaccination are minuscule.

But remember: Motivated reasoning is incredibly powerful. We interpret facts in a way that supports some underlying belief we already hold. Many people whom we would describe as vaccine-hesitant in fact harbor negative beliefs about vaccines. Motivated reasoning will seek to confirm those beliefs, as refuting an underlying belief requires cognitive change, which takes effort. This is why many vaccine-hesitant individuals can look at the one in ten thousand risk of heart inflammation and compare it to a one in one hundred risk of death and conclude that the vaccine is not safe enough.

I've seen this type of reasoning many times. Often, it requires two steps of data misinterpretation. The first inflates the vaccine risk: "Probably the numbers are higher than one in ten thousand, but cases are being missed." And the second diminishes the risk of the virus: "I'm healthy, and I will be fine if I get infected" or "The death risk is being inflated somehow." The effects of this reasoning lead to the conclusion that was inevitable all along: "I won't get the shot."

Not all vaccine-hesitant individuals are armchair epidemiologists, looking up VAERS data, but many are still engaging in the same cognitive processes, often using anecdotal data (a friend of a friend had a bad vaccine reaction) to justify their underlying belief that the vaccine is too risky. And anecdotes abound in the misinformation age—your social media page is full of them, and the algorithms that drive the social media feed will ensure that you see more of them if you engage with that content. The effect is that those who are on the fence with regard to vaccination are exposed to constant data points that support the conclusion that, deep down, they want to be true: that they don't need to get the vaccine.

While a robotic appraisal of the facts should be enough to convince any adult that the risk of vaccination is far less than the risk of COVID, we are not robots. Instead of examining facts and reaching conclusions, we reach a conclusion and go on the search for facts. The public is not alone in this. Doctors do it too. And you need to know how to recognize it when they do.

When Your Doctor Wants a Conclusion to Be True

Doctors engage in motivated reasoning in ways that can be hard for patients to pick up on. Most of us have three main drives when it comes to patient care: to make people better, to not make people worse, and to avoid needing to deal with the insurance company. I'll go into greater detail on insurance companies in chapter 10. Here, I want to address how the fact that we want our patients to be well can strongly color how we interpret the objective facts about their cases. In brief, our motivated reasoning can lead us to downplay the seriousness of your condition.

This effect has been most robustly described in the cancer literature. When a patient has a terminal cancer diagnosis, they will often ask their physician how long they have to live. It's a terrible conversation to have, and the data suggests it might not even be worth it. A synthesis of twelve studies compared doctors' predictions about patients' survival to reality. Doctors' predictions were almost always overly optimistic, typically giving the patient 50–90 percent more "time" than they would turn out to have.

How do we justify those overly optimistic predictions? We interpret the data to get to the conclusion we want. We may look at a low white blood cell count and blame it on the chemotherapy instead of the cancer infiltrating the bone marrow. We may shrug off a patient's complaint of severe fatigue as the result of a restless night's sleep, because the alternative—that the cancer has recurred—is something we simply don't want to be true.

This won't be fixed by using the old "Give it to me straight, doc." We aren't lying or sugarcoating. We are misinterpreting data to get to a conclusion we want to be true, just like you do. Just like every human does. And it should be obvious how this optimism may, paradoxically, compromise your health. It may lead your doctors to be less aggressive, or to forgo a test that you actually need, because the conclusion we want to reach—that you are healthy—may be the wrong conclusion.

So how do you know whether your doctor is processing information properly? One powerful tool that has emerged in this space is called the "'surprise' question." Instead of asking a physician how long a patient has to live, we can ask "Would you be surprised if this patient lived six months?" "Nine months?" "Twelve months?" And so on. These simple, binary choices help to short-circuit the motivated reasoning that pushes us in overly optimistic directions. And the "surprise" question isn't limited to predicting death. It can be used to get an honest assessment of the likelihood of response to a medication or other treatment, the risk of surgery, or whether a lifestyle intervention is likely to meaningfully improve your health. If you ask a doctor whether taking ginkgo biloba will improve your cognition, they are likely to reply "Maybe. It couldn't hurt to try." But if you ask "Would you be surprised if taking ginkgo biloba improved my cognition?" you're much more likely to get the more realistic "Yes, I would be very surprised." Changing the script a bit can make all the difference in these types of medical interactions.

Once you have your sensor tuned to it, you'll start seeing motivated reasoning everywhere—in the politicians on the news, in your own friends and family, in your doctor, and, hopefully, in yourself. Recognizing it is only the first step, though. You need to know how to fight against it. You need to *de*motivate your reasoning.

Four Ways to Demotivate Our Reasoning

Because motivated reasoning involves... well, reason, we might not notice that we are using it unless we are highly introspective. We may think we are being perfectly logical. But there are three rules of thumb I've developed that help me figure out if I might be in a motivated-reasoning situation. I'm hopeful they will help you as well.

One thing to remember is that you should never draw a conclusion from a single piece of data, no matter how compelling. The problem is that humans are conclusion-drawing machines. Our brains are simply wired that way. As soon as our brains are developed enough to form conclusions, we start jumping to them. One of my first memories of school is from Mrs. Wimble's nursery school class. One day, we played a game where each of us reached into a dark box to feel something inside. Without looking, we had to tell the class what it was. When I reached into that box, I confidently blurted, "It's an orange hair clip." It was a white hair clip. I was too young to know that the picture my mind had created of the thing in the box was not necessarily accurate. And while I would not make that mistake again, the same neural pathways are active in all of us—leading us to feel confident in our conclusions because they comport with the way we want, or need, or expect the world to be.

The first way to avoid motivated reasoning is to be thorough. Look for more data. Avoid the temptation to draw conclusions no matter how *right* it feels that the hair clip is orange. If you hear that a vaccine led someone to speak in tongues, wait for more data. If you hear that a new virus is turning people into zombies, wait for more data. True conclusions are supported by multiple lines of evidence.

This orange hair-clip phenomenon happens to doctors all the time. As a kidney specialist, I am often asked to perform dialysis on patients in the hospital. The typical situation is a patient whose kidneys have started to fail due to infection, low blood pressure, or complications of surgery. Dialysis is a procedure where that patient's blood is circulated through a dialysis machine, which can function as an external kidney—filtering out the toxins and other metabolic detritus before returning clean blood to the patient. One question that comes up when we consider this treatment is: When should we start? Too early, and we might end up treating people who would recover on their own, putting them at risk of dialysis complications (like blood clots, infection, and pain) for no reason. Too late, and we might miss the window to really help them.

So how do we know how high we should let the toxin levels get before we hook a patient into the machine? In May 2016, the results of a trial of early versus late dialysis in a single hospital appeared in the *Journal of the American Medical Association* (*JAMA*). The conclusion: Early starts saved lives. As someone who provided dialysis therapy, this conclusion *felt* right to me—the thing I did to help patients really did help. But two months later, a larger study appeared in the *New England Journal of Medicine* evaluating the same question across multiple different hospitals. The conclusion: An early start versus a later start made no difference. The conclusion I would *like* to reach, that dialysis is helpful early in the disease course, was supported by one trial and refuted by another.

By being aware of my own motivated reasoning, I was able to force myself to look for other data sources and then adapt my thinking on the issue. By expanding the number of data sources I considered, I was better able to avoid cherry-picking studies to support my personal preferences. My current practice, based on the best available data, is to delay dialysis if possible to try to allow patients to recover on their own, but to use it when needed as a last resort.

The second way to avoid motivated reasoning is to realize that, in Medicine as in life, there are no guarantees. Health can feel chaotic and random. While we can increase our chances of leading long, healthy lives by exercising, eating well, and not smoking, we all know someone who was the very picture of a healthy lifestyle and yet got struck down by cancer, heart disease, or infection. A virus that can be asymptomatic in some people and fatal in others, as is the case for COVID-19, is terrifying on its face. The conclusion "An invisible bundle of nucleic acid may kill me in the next few months" is untenable, and so we all use motivated reasoning to find some other conclusion that we can live with. More acceptable conclusions range from "I am healthy enough that if I get COVID-19, I'll be fine" to "The cure for COVID-19 is in the vitamin aisle" or "COVID-19 doesn't even exist." I have heard them all, they are all comforting, and they are all wrong. The truth is, unfortunately, we are not in total control of our health—we can only hedge our bets.

Nevertheless, the *promise* of control is a powerful one, and it relies heavily on motivated reasoning. Entire industries exist around this promise, and they are ripe for fraudsters and hucksters. Remember *Natural Cures "They" Don't Want You to Know About*? This was the first in a series of books written by convicted fraudster Kevin Trudeau, which purported to provide simple cures for basically every modern medical ailment from depression to cancer. These types of books appeal to our motivated reasoning because they tie directly into the conclusion we want to be true—we *want* a cure. The unspecified "They" in the title can certainly be read as "Doctors" if you are so inclined, which creates a narrative: Doctors don't want you to know the cure because then the doctors would not make money.

Books like these are designed to erode trust in Medicine, in this case to convince readers to give more money to the author. (The book was more or less an advertisement to subscribe to Trudeau's newsletter.) Once we are primed to think about the conclusion we want to reach, we are highly suggestible to even the most dubious facts. And the facts in Trudeau's books were *really* dubious—including the claim that AIDS is a hoax designed to sell HIV medications. The book series sold snake oil as a guarantee of a healthy life. But however much we want them, there are no guarantees.

You may feel that a treatment has to work because the alternative is too painful to consider, but this is just reasoning to a conclusion you want to be true. In reality, we know that the treatment may work, and it may not. A good doctor will help you play the odds to pick the treatment with the best chance of success—but that chance will always be less than 100 percent. If you find yourself believing otherwise, you are likely engaged in motivated reasoning.

The third way to avoid motivated reasoning is to adopt the strategy of a Fortune 500 company and outsource. It can be hard to recognize motivated reasoning in yourself, but it is actually fairly easy to recognize in others. I'm sure you've had a friend or relative tell you how they just know they'll strike it rich in that multilevel marketing company or that a certain diet will help them shed those pounds, and you smile and nod, and think, *Sure, good luck with that.* Well, you can use your friends and family to your advantage when you are making medical decisions. Bring others into your circle, share the facts with them, and then (and this is the hard part) *listen* to what they say, even if their conclusions don't agree with yours. *Especially* if their conclusions don't agree with yours.

Spouses are particularly good in this role, I've learned. I remember sitting in the clinic at the Veterans Affairs hospital in West Haven, seeing a man with diabetes and kidney disease. We had tried to control his blood sugar levels using pills for six months, and they weren't working. His sugar was too high, and it was making his kidney function worse. I advised him that we needed to start insulin injections. This was a conclusion he did not want to hear—the idea of multiple injections along with the finger sticks to manage dosing seemed overwhelming. He told me he could control his diabetes with diet, and I agreed that dietary changes might be helpful but probably not enough to avoid needing some insulin. He suggested we increase the doses of the pills he was taking. I told him he was already on the maximum recommended dose. He suggested that his kidney function would be fine, that we could keep a close eye on it and if things got worse, *then* he would do the insulin. It was then that his wife, who, fortunately, had come to the visit with him, called him out. "Listen to the man, George. He's trying to help you." And I was. George, with the loving support of his wife, integrated insulin into his diabetes care. He's doing well today.

So far, we've addressed three main ways to avoid motivated reasoning: Wait for more data. Be suspicious of guarantees. And outsource the job to others. But the most important way to avoid this, the overarching way, is to know yourself—to be mindful of your own worldview, what you need to be true, and what you want to be true.

This metacognition, thinking about thinking, can be tiring. Introspection is work. But it is critical. When we make a decision, we need to ask ourselves *why* we made it. And we need to be honest about the answer. Was it really a rational appraisal of the facts at hand, or maybe, just maybe, could it be that you used the facts to justify the decision you were going to make anyway? Did you decide not to take that medication because you think the risks outweigh the benefits, or did you come to that conclusion because you don't like the idea of taking pills? Or because you don't like the idea of not being perfectly healthy? Or because it makes you feel like you failed somehow—not being able to maintain your body in its natural state? Or because it makes you feel old? You need to ask yourself these questions and be open to the real, honest answers.

While turning on your motivated-reasoning sensor will help you make better medical decisions, it's actually much bigger than that. You'll begin to see how deeply we all want to avoid change, to continue in the habits and beliefs we've held for so long. And by seeing that, you can begin to change yourself. You can open yourself up not only to new evidence-based medical therapies but to new experiences, new beliefs, and new ways of thinking. Importantly, you can open yourself to trusting your doctors, realizing that we care much more about taking care of you than taking care of the conclusion you want to be true.

There's a reason this topic is addressed in the first chapter of this book: because recognizing our own tendency toward motivated reasoning is the first step toward opening our minds to the truth that is out there. The rest of the book will show you how medical truth gets discovered and teach you how to recognize it when you see it. Now you know how to treat truth fairly.

CHAPTER 2

Changing Our Minds

IT WAS A cold December night in 2007 and I, in my second year of internal medicine residency, was on my own for an overnight shift in the intensive care unit. A perk of second year was that I was put in charge of half of our forty-bed ICU without in-person supervision. Of course, there was an attending physician I could call if I got into trouble. But he was at home sleeping. In the moment, all eyes would look to me.

It was exhilarating. Patients would come up in a steady stream from the emergency room or the hospital floors, all needing immediate, aggressive medical attention. One patient with uncontrollable bleeding from the gastrointestinal tract, another with worsening lung disease requiring mechanical ventilation, another who suffered a heart attack needing to be stabilized before he could get surgery, another with liver failure from a Tylenol overdose. Younger then, I enjoyed the challenge. (Today I'm not sure I could do it for sixteen hours straight.) There was rarely any downtime; I would run from room to room, feeling like one of those old-time entertainers keeping plates spinning atop poles, desperately trying to keep them all from breaking. Just as you get one plate stabilized, another starts to wobble, and so on.

In the morning, the day team, the attending physician, fellow, and multiple residents and students would arrive. The cavalry. I would make the report of what happened overnight with casual pride: "Yeah, I kept everyone alive all night. No big deal." There were congratulations all around, and I would head home, grab a cheesesteak on the way, and fall asleep on my couch with *Star Trek* reruns on the TV. The next night, I would do it all again.

The cases were varied in the ICU, but you couldn't go a night without seeing at least one case of septic shock. Septic shock occurs in the setting of overwhelming infection. The blood vessels, normally taut and springy, are poisoned by the toxins floating around the body and begin to dilate and deflate. As that happens, the blood pressure drops, decreasing oxygen delivery to the vital organs, which begin to fail. The heart, trying to compensate, races—sometimes so fast that it begins to fail as well. It's a vicious cycle, and even with the best antibiotics, one that is not easily reversed once it has begun.

In those days, the treatment of septic shock was dominated by something called "early goal-directed therapy," or EGDT. The idea, popularized by critical care specialist Emanuel Rivers in a 2001 *New England Journal of Medicine* paper, was powerful: Treat septic shock with specific targets in mind. The Rivers algorithm could be easily memorized—I can still recite it. First, get the central venous pressure (monitored with a special catheter) above 8 using IV fluids. After that is achieved, check the blood pressure. If the mean pressure is below 65, give vasoconstrictors, like adrenaline, until the pressure is higher. Once that is done, measure the venous blood oxygen levels, and increase them by either giving blood transfusions or medicines that strengthen the heart's pumping.

The data from Rivers's paper was clear: 47 percent of patients with septic shock who had been randomized to the usual care died, compared to 31 percent randomized to his early goal-directed therapy algorithm. EGDT had thus become the new standard of care. There was something so nice about this protocol. It made treating septic shock feel like flying a plane. Are the flaps up? Check. Is the engine hot? Check. It was reassuring—do *this*, the protocol said, and your patient will be okay. And if they are not okay, at least you've done everything you can.

The first step of the protocol—give fluids until the central venous pressure is above 8—became something of a contest of machismo among the residents. Under the belief that everyone is always too wimpy with fluids, we would often brag on morning rounds about just how much fluid we gave our patient overnight. Six liters? Ten liters? The higher the number, the more nods of approval you got.

One night I gave a twenty-year-old patient with cancer and septic shock twelve liters of fluids in twelve hours.

That was the wrong move.

At the time, I didn't know it. None of us did. In fact, when I told the morning team what I had done, they praised me for my quality doctoring. I had followed the protocol to a T, hanging bag after bag of IV fluids. The team thought my efforts to save this young life were . . . well, heroic.

The patient died. Not that day. But over the next several weeks, organ system after organ system failed. We did our best—treating new infections as they arose, adjusting the ventilator settings to increase blood oxygen levels, keeping the blood pressure up with adrenaline. But he couldn't be saved.

We don't often tell families this, but we usually know a patient in the ICU will die weeks before it happens. We don't tell in part out of fear of their response, in part out of optimism, in part out of a vague superstition that admitting what will likely happen will bring it about. But also because it's hard to put our finger on exactly *how* we know. You just get a sense: The blood pressure stops responding to the adrenaline. The skin and abdomen fill with excess fluids. Often what little interactivity the patient had is lost as they slip deeper and deeper into this liminal peri-death.

When we see this occurring, there is very little we can do to change the outcome. We can often keep people alive using machines and medications, but in reality the patient is not recovering, just treading water. My patient, to whom I had heroically given twelve liters of fluids on that first night, died after a long ICU stay. His body was dependent entirely on machines until, at the family's request, we turned them off.

I did not feel guilty. Patients die in the ICU all the time. We do our best. We save some, we lose some, and we move on. And, I told myself again and again, I had done *exactly* the right thing. I had followed the protocol—the best available evidence. Sometimes, you just lose.

A few years later, the first papers questioning the aggressive use of intravenous fluids in septic shock started coming out. While it wasn't a direct backlash to the Rivers protocol, the medical establishment had started to realize that our new love affair with fluids in septic shock was problematic. Flooding patients' bodies with saline the moment they came into the ICU left you with a problem: That fluid had to go somewhere. The extra fluid puts extra stress on the heart, which had to pump it around, and the lungs, which had to fight to remain clear. In 2017, sixteen years after the Rivers paper was published, the *New England Journal* published the combined results of three early goal-directed therapy trials. In contrast to the 263 patients enrolled in the original study, these studies had a combined 3,723 patients. And the finding was different. The death rate among those randomized to the Rivers protocol was the same as it was among those randomized to conventional, nontarget-based care—around 25 percent. And those who got the Rivers protocol spent more time in the ICU, were more likely to require heart support, and had higher hospital bills. In other words, for about a decade, we had been wrong.

The best way I have to understand this series of events is to compare it to a pendulum. Before Rivers, we were probably giving too little fluid in the ICU. After, we gave way too much. Now, we are somewhere in the middle. Many people think medical knowledge is permanent and unchanging, and for some things, that's true. We have known for ages that the brain has four major lobes. That is unlikely to change. But our understanding of what happens within and between those lobes is constantly evolving. We have known for ages that the heart circulates blood around the body and that the kidneys filter that blood to produce urine, but we are continuing to learn how the heart and kidneys signal their needs to each another and respond to each other.

The issue is that you don't often hear news stories telling you that the brain still has four lobes or that the heart still circulates blood. Medical news and medical discovery lives on the edge of understanding. What you hear about is *new* knowledge—and new knowledge is not nearly as stable as old knowledge. By nature, it has had less time to be replicated and confirmed. This is why some people think Medicine is ephemeral, driven by special interests rather than science and, therefore, untrustworthy. In reality, the process of discovery is messy.

In 2007, I didn't know that the *New England Journal* would effectively end the practice of early goal-directed therapy ten years later. All I had to go on was Rivers. But I find myself thinking back to that patient frequently, knowing that if I had to do it all over again, I would have given far fewer fluids (and certainly not bragged about it). What I did was not malpractice. Ironically, failing to do what I did might have been considered malpractice at the time. But it was wrong, the type of wrong born of ignorance as opposed to malice.

Still, I find myself feeling thankful that we operate in a system that has the capacity for self-correction. If Rivers had been taken as gospel, unquestionable and unassailable, more people would have died. How much better that we have a tradition of questioning in Medicine, a tradition of rigorous study that can prove that what everyone believes is best isn't always. A tradition, in short, of changing our minds.

How Minds Change

When a doctor changes their mind, patients often lose faith in the doctor. The expectation, however unrealistic, is that we are paragons of knowledge, at least within our domains, and that changing course in the path to diagnosis and treatment shows that, underneath the white coat, we are merely human. Understanding how minds get changed in general and how doctors change their minds in particular should, I hope, show that flexibility of thinking is not only healthy; it is necessary. In other words, a doctor changing their mind is a good sign—a sign that they are thinking, examining new data, continuing to work for you. The best doctors are, at heart, scientists, not priests. Their knowledge is earthly and imperfect and evolving. And that's okay. It is how we all move forward.

In the last chapter, we focused on motivated reasoning—using the amazing powers of our brain to twist facts until they suit a conclusion we were hoping for all along. Motivated reasoning happens when we already have strong feelings about what we believe to be true or what we want to be true. And when it comes to our health, there's nothing we want to be true more than to believe we are, or can become, healthy. But we don't feel that strongly about *all* our beliefs. Indeed, we change our minds all the time, from what we want to eat for dinner to what outfit we're going to wear to work. When it comes to our health, we change our minds frequently too, moving from one diet to another, trying new exercises, taking vitamins, stopping taking vitamins, and so on.

But we aren't just blindly flitting from one belief to another, fickle dilettantes riding whims and fancies. In fact, there is fairly robust scientific literature that explores why people change their minds, and it shows that the ability to change your mind is critical. It means you are being flexible and open to new information. And for many patients, the first step toward healing is changing their mind about something—whether it is a decision to stop smoking, increase physical exertion, or agree to the chemotherapy they were afraid to start. But you can change your mind for good reasons and for bad reasons, and distinguishing between the two can make all the difference.

One of the simplest reasons people change their minds is because they are forced to acknowledge a new reality. In 2018, psychology professor Kristin Laurin published a study in the journal *Psychological Science* examining how San Franciscans felt before and after the city introduced a plastic water bottle ban. Before the ban, most people viewed it somewhat unfavorably. Days after the ban, the favorability shot up. She replicated this finding before and after a smoking ban in Ontario, Canada, and before and after the inauguration of Donald Trump in the United States. The theme was clear throughout—we find things more acceptable once they have actually happened than we do when we are anticipating them happening.

Changing your mind to adapt to a new situation (over which you have little or no control) is almost certainly a positive thing—it keeps us from falling into despair. In the context of health, it is quite important. Patients confronting a new diagnosis of cancer or dementia or heart disease almost universally find they are *less* distraught than they thought they would be beforehand. We adapt quickly to the new normal, shifting our focus from the things we can't change to the things we can.

Our minds can also be, for lack of a better term, beaten into submission—changed by being inundated with the same information over and over and over again. And it doesn't matter if the information is true or not. This phenomenon is known in psychological circles as the "illusory truth effect": People believe what they hear if they hear it enough.

First described in 1977, the effect was documented by (as usual) exposing psychology students to a carefully constructed test. In this case, they were presented with sixty statements that all seemed plausible but were not necessarily true—for example, "Cairo, Egypt, has a larger population than Chicago, Illinois." They had to rate how true each statement was. The test was completed three times two weeks apart. In the last two sessions, some of the statements were replaced, but twenty statements appeared on all three tests. The researchers found that the statements that were repeated were rated more likely to be true when they were seen a second time (regardless of whether they were *actually* true). Familiarity had bred trust.

Illusory truth is used across all domains of society, from advertising, to news, to, of course, politics. Donald Trump, ever a marketing whiz, used this to great effect with his well-documented nicknames for political rivals. From April 2016 to October 2020, his tweets included the phrase "Crooked Hillary" more than 350 times.

Social media bombards us with illusory truth because social media algorithms are more likely to surface information similar to information you have interacted with before. Prior to social media, the claim that a pizza place in Washington, DC, was the epicenter of a child sex-trafficking ring would have found little purchase. Social media, by surfacing the same false statement about Comet Ping Pong pizzeria over and over again (to people inclined to interact with those statements) made people believe it was true—the sheer repetition lent it credibility.

This phenomenon is a corruption of brain circuitry that served us well for the majority of human history. If you are part of a small community or tribe—as our ancestors were—believing what you heard multiple times made sense. If everyone in the village tells you not to go near the cave with the bones outside it because a dangerous creature lives there, you would do well to believe them. In the modern era, though, our village is no longer the people who happen to live around us. Social media can create microvillages, or bubbles, of people from around the world who share the same beliefs (false or true), which reinforces those beliefs inside the bubble. Breaking out of the social media bubble is uncomfortable, because it forces us to interact with beliefs at odds with our own. Remember, motivated reasoning defends the status quo. And change is hard when our beliefs are strong.

The best way we can change our minds is by updating our beliefs as new information becomes available. This is easiest when those beliefs are not firmly held, deeply entrenched, or emotionally resonant. For example, my kids really like chocolate. They particularly like chocolate after dinner, just before bedtime. My wife had the strong belief that giving them said chocolate right before bedtime led to "craziness" and should be avoided. I held the belief that our kids were always crazy and it had nothing to do with the chocolate. Life went on, and the kids kept getting chocolate (except sometimes when they got something else, or the chocolate ran out). And my beliefs began to change. I kept seeing evidence of the chocolate-craziness connection, night after night. Chocolate, crazy, chocolate, crazy. Each time that link was demonstrated, my belief in my wife's hypothesis strengthened, and my belief in my own hypothesis weakened. Eventually, the weight of evidence was too much to bear, and I conceded she was probably right. These kids *are* crazy, but they are extra-crazy with chocolate in their systems.

Allowing the slow, methodical updating of your prior beliefs based on new data is a form of Bayesian reasoning. Thomas Bayes was an eighteenth-century English minister and statistician whose eponymous theorem demonstrated, mathematically, how new information can be used to update prior knowledge. It is a centerpiece of statistics to this day, and the bane of many undergraduate stats majors. But Bayesian reasoning is quite similar to how the human mind works, at least when we are not unreasonably tethered to conclusions we want or need to be true. We have a belief, we get new data, and we update our belief.

Imagine that you see a suitcase lying on the side of the road. You may believe it is full of clothes. But then new data arrives: Someone comes to pick it up and struggles, straining every muscle, and then begins dragging it along the sidewalk. You use that data to update your prior belief; it's probably not clothes in the suitcase after all.

This Bayesian reasoning is how we need to approach all information we receive. We shouldn't simply believe whatever new piece of data comes into our brains—we should evaluate it in the context of what we already knew and what we already believed, and then, critically, we should *allow* that data to update our beliefs. Note how this differs from the process of examining new data when motivated reasoning is in play. In that circumstance, instead of allowing new data to adjust our understanding, we reject new data that doesn't fit our understanding. Bayesian thinking forces us to grapple with new data, whether we like it or not.

If this sounds a bit like the scientific method, it's because it is. At least when the scientific method works correctly. (In later chapters, I'll show you how the scientific method can go awry.) When it works, the scientific method is an iterative, stepwise process that inches us closer and closer to the truth about the world. Emanuel Rivers wrote his paper, and our beliefs about how to treat septic shock changed based on the data he showed us. Subsequent data came out that led us to revise those beliefs. In the end, we're treating sepsis better now than we ever did in the past—all because we were willing to change our minds. If we retain that flexibility, I am confident we'll be able to treat sepsis even better in the future. Medicine always has room for improvement.

Where Doctors Go Wrong

Doctors make mistakes, and mistakes erode trust. When patients think of medical errors, though, they tend to think of errors of commission—giving the wrong medication, operating on the wrong body part. Errors of omission—neglecting the appropriate diagnostic test or missing the key diagnosis—are far more common, harder to identify, and probably more deadly. In the Institute of Medicine's seminal report *To Err Is Human*, researchers estimated that up to ninety-eight thousand Americans per year die of medical errors, similar to the number who die each year of Alzheimer's disease.

A doctor who tells you they have never made an error is either enjoying their first day on the job or lying. I remember a particular case during training where a colleague had written that the dose of a medication be given "IV" instead of "IM," intravenously instead of intramuscularly, as it was intended. We discovered the mistake quickly but too late—the dose had already been given. We monitored the patient intensively for twenty-four hours, waiting to see if anything bad would happen. Nothing did, fortunately. But that was luck. The situation could just as easily have been disastrous.

Errors in the moment are devastating to physicians, and to the bond of trust that physicians have with patients and society. We tend to ruminate on them, often for years after the fact. We hold morbidity and mortality conferences, to discuss our errors in an unbiased environment, and perform root cause analysis, to try to ensure that they don't happen again. But this error-correction system misses another kind of mistake—a quasi-mistake, really—that we rarely reflect upon and that doesn't get captured in national statistics. Nevertheless, it is likely the most common mistake doctors make.

As in the treatment of my young patient with septic shock, we do the wrong thing because medical science doesn't know better at the time. Is it any wonder why patients get frustrated with doctors when they feel that we are constantly changing our minds? If you had a loved one in the ICU in 2004, your doctor would have confidently told you that they administered "aggressive fluid resuscitation." If another loved one was admitted today and you asked the doctor when the aggressive fluid resuscitation would begin, they would tell you, "Well, it won't."

This *feels* fickle. Examples abound of well-informed doctors changing their tunes. In the 1980s, if you asked a doctor how to lose weight, they would tell you to avoid eating fat. Today, they would be much more likely to tell you to avoid eating carbs. During the coronavirus pandemic, Anthony Fauci, director of the National Institute of Allergy and Infectious Diseases, was pilloried in the right-wing media for his changeable stance on mask wearing. Early in the pandemic, when hand-to-hand transmission was thought to be the main infection route, Fauci had argued that masks were not helpful. Later, when droplet transmission was shown to be the predominant route, he acknowledged that he had been wrong.

It is frustrating—frightening, even—to realize you were wrong, but it is necessary. What happened to me (and to Fauci) was scientific progress. Science is the *process* of discovering truth, and it is imperfect. Individual studies may lead us down unproductive or even harmful paths, but the use of the scientific method has transformed Medicine from an esoteric, quasi-mystical art to its current form, which, to our predecessors, would appear miraculous.

While flip-flopping on medical advice can have the effect of reducing public trust in doctors, we should be more concerned about a dogmatic adherence to a principle or idea. This is an area where certain types of alternative medicine thrive, often offering history as a proxy for truth. One could ask how acupuncture has evolved over the centuries, and many acupuncturists will proudly point out that it hasn't and that that is a strength. The same can be said for some herbal medicine and homeopathy. It seems almost a tenet of practicing these arts that they cannot be questioned, whereas mainstream medicine is under nearly constant scrutiny. The side effect? Docs like me are sometimes wrong. Alternative medicine practitioners are often wrong too, but you (and they) may never actually find out.

The progress of science is not a straight line from ignorance to truth, populated by stepping-stones of exquisitely designed studies. Progress is more of a meandering path, filled with cul-de-sacs and dead ends, areas where we have pursued truth and found none. The practice of phrenology, reading the bumps on people's heads to determine their personality, for instance, was extensively researched (albeit using nineteenth-century research techniques), and yet we can conclude at this point that there was no deeper underlying truth being discovered there. That well was dry.

So how do medical researchers know where to dig for truth? How do we decide what to study? And how can patients know whether what we are studying has any chance of becoming useful?

The Double-Edged Sword of Biologic Plausibility

There are infinite medical questions we could formally study. Do jumping jacks prevent colon cancer? Does eating celery reduce the risk of falls? We could design rigorous studies to answer those questions, studies that could change the minds of the medical establishment, but we don't—because it's likely that the answer to those questions is no. Could we be missing something here? Of course, but there are only so many humans in the world and only a fraction of them are medical researchers; we have to go after things with a reasonable chance of success.

The chance of success in a medical study rests on the concept of "biologic plausibility." Is it plausible, within our current understanding of how the human body works, that eating celery reduces the risk of falls? Not really. Is it plausible that lower-body strength training reduces the risk of falls? Definitely. So we start by studying that. Regardless of the results of a medical study, if the underlying hypothesis being tested is not biologically plausible, you should be very skeptical. For example, it is not biologically plausible that putting a red jelly bean on top of your left knee will cure your headache. If you come across a rigorous study that concluded that, in fact, red jelly beans do have a profound effect on reducing headaches when placed on the left knee, you should be highly suspicious that you aren't getting the whole picture. Was the study perhaps funded by "Big Jelly Bean"?

However, just because a treatment is biologically plausible does not mean it will work. Failure to appreciate that simple fact has led to a lot of failed experiments. This is because biologic plausibility can be too flexible. Completely opposite things can *both* be biologically plausible.

For example, SARS-CoV-2, the virus that causes COVID-19, binds to a receptor on the surface of cells, called the "ACE-2 receptor." That "ACE" in the receptor name is the same "ACE" as in "ACE inhibitor," a class of widely available and cheap blood-pressure drugs. Could ACE inhibitors stop the spread of SARS-CoV-2? ACE inhibitors bind to that ACE-2 receptor, which would block the virus from attaching, decreasing the risk of infection. It is therefore biologically plausible that giving people ACE inhibitors will reduce the risk of COVID. But wait. The use of ACE inhibitors also leads your cells to produce more ACE-2 receptors as your body tries to compensate for the receptors that have been blocked. So maybe giving people ACE inhibitors will worsen the severity of COVID-19 infection. Two opposite hypotheses: ACE inhibitors will help, or ACE inhibitors will hurt. Both are biologically plausible.

This is why I remind my trainees time and time again that biologic plausibility is the *start* of medical research, not the end. One of the central innovations of modern medical science, in fact, was a move away from blind trust of biologic plausibility to empiric testing of hypotheses. This was made possible, in large part, due to the work of statistician Ronald Fisher in the 1920s, who figured out the math that would allow rigorous comparison of two populations of individuals receiving different treatments. Before Fisher, doctors went with their personal experience, with their gut, or with whatever the most important doctor in the room told them to do. Enabled by Fisher, medical research would become ever more reliant on hard data and statistics, culminating in the evidence-based-medicine movement that thrives today.

Today, we start with a biologically plausible idea, but we don't stop there. We test it. At the time of this writing, there have been multiple studies looking at the link between ACE inhibitor use and severity of COVID-19. So far, the results are basically null: They don't hurt; they don't help.

Biologically plausible ideas that have *not* been rigorously tested form fertile ground for medical errors, and also lucrative businesses. Over the past few years, for example, stem cell therapy clinics have sprung up all over the country, offering treatment for a variety of ailments, ranging from arthritis pain to thinning hair. The clinics take a sample of your blood and put it through a machine to separate out stem cells (cells that have the capacity to grow and divide into a variety of tissue types). The theory goes that injecting such cells into an arthritic joint, for example, might promote the formation of cartilage, reducing pain. Is this theory biologically plausible? Sure. One stem cell clinic's website proudly states that it will "harness your body's natural healing ability." It's big business—an estimated $10.2 billion market in 2021. And those costs are almost entirely out-of-pocket for patients.

Yet this therapy has never been formally tested. The entire multibillion-dollar industry is founded solely on biologic plausibility. And with all that money at stake, who would agree to test it? Best-case scenario: The practitioners prove it works, and they keep making money. Worst-case scenario: They prove it is no better than a fancy placebo, and the whole thing shuts down.

I take a firm stance that biologically plausible treatments still need to be rigorously tested before wide adoption. But plenty of people argue against this, particularly when the intervention is relatively low-risk and the mechanism of benefit obvious. Sometimes, the argument boils down to one oft-repeated analogy about parachutes. It goes like this: "Would you do a randomized trial to test if a parachute will save your life when you're falling out of a plane? I sure as hell wouldn't want to get the placebo parachute!"

This argument is, in my opinion, somewhat tortured, but it gets brought up all the time in the medical literature. In fact, a 2018 journal article by oncologist Vinay Prasad and his team cataloged dozens of medical articles where a therapy was likened to a parachute. Among those were articles suggesting that using simulation-based training for medical professionals was obviously beneficial to patients and didn't need to be tested; that in vitro fertilization was so obviously beneficial for achieving pregnancy that it didn't need to be tested; and that treating diabetes with agents that lower blood sugar levels was so obviously beneficial that it didn't need to be tested.

All of these therapies make sense. It is definitely biologically plausible that reducing blood sugar would be good for people with diabetes. But interestingly, twenty-two of these medical "parachutes" went on to have formal randomized trials. And in only six of those cases did the "parachute" prove to be effective. In five cases, the parachute was worse than the alternative. The remainder had mixed results.

Remember, it was doctors and scientists who were describing these interventions as "parachutes," in some cases arguing that testing them would be not only wasteful, but also unethical. And yet we often find that when tested, our assumptions—grounded in centuries of medical knowledge—are wrong.

For patients, the conclusion should be pretty straightforward: Don't start an intervention or take a medication or supplement because it makes sense that it *should* work. Ask if it has been proven (ideally, in the context of a randomized trial) that it *does* work. If not, you're likely to get burned.

The Obviously Beneficial Vitamin E

We've been burned before. Such was the case with vitamin E. Vitamin E is actually the name for eight related compounds that are produced by plants, probably to protect the fat in their seeds from going rancid in the presence of oxygen. This antioxidant effect is quite powerful. In fact, the primary role of vitamin E in humans seems to be to protect our cell membranes (filled with fats) from being damaged by all the oxygen and oxygen byproducts that result from us . . . well, breathing.

It is almost impossible to become deficient in vitamin E, as it is present in basically any food that contains fat. Hence, there is no disease of lack of vitamin E the way we have scurvy for lack of vitamin C, rickets for lack of vitamin D, or beriberi for thiamine deficiency. True, vitamin E deficiency can happen due to some rare genetic conditions and from syndromes of fat malabsorption—but chances are, if you're reading this, your cells have all the vitamin E they need.

Nevertheless, there is something compelling about that antioxidant idea. There is plenty of research showing that oxygen byproducts—free radicals—damage cells. This oxidation damage has been implicated in a slew of diseases, including heart disease and cancer, and even in the aging process itself. Vitamin E scavenges up those free radicals to protect your cells. So it is biologically plausible, then, that boosting your vitamin E levels would suppress this damage. Moreover, vitamin E is safe at normal dosages, cheap, and widely available.

At this point, you may well be thinking it would make sense to run out and buy some vitamin E tablets. It won't be hard. An entire industry exists extolling us to part with our hard-earned cash for a capsule filled with vitamin E. But remember, all this biologic plausibility should be just the beginning of the research process.

Fortunately, vitamin E supplementation has been formally tested. Across multiple studies and multiple decades, participants were randomly assigned to take vitamin E or a placebo and followed for the development of a host of diseases vitamin E was thought to prevent. Overall, no effect. Vitamin E pills didn't protect against heart disease, cancer, or aging. They didn't make people's skin softer or more wrinkle-free. In fact, one major study showed a higher risk of heart failure in the group treated with vitamin E compared to the group who got a placebo.

If your doctor was paying attention throughout this period of discovery, they may well have started out telling you it wouldn't hurt to take a vitamin E pill once a day and ended up telling you to throw them in the garbage. This is often read as being wishy-washy. It is most certainly not. Dogged adherence to a belief that turns out to be wrong is cowardice. When your doctor tells you they have learned new information and are changing their recommendation for you, it's a really good sign that you are under excellent care.

Blind faith in biologic plausibility can lead doctors and patients to make decisions that turn out, in the future, to be the wrong ones, costing you money for something that isn't benefiting you. But what if a biologically plausible treatment survives formal testing through the rigors of a randomized trial? Are we in the clear to embrace the treatment?

Not necessarily, because biologic plausibility has a close relative that, even *after* a rigorous trial, may lead us to believe a treatment works when, in fact, it doesn't.

"Close Enough" Outcomes

Surrogate outcomes are a research tool closely related to the concept of biologic plausibility. I'll give you the formal definition of a "surrogate outcome" in a moment, but basically it is something we can measure easily that is considered close enough to a truly important outcome, which may be harder to measure or rarer. As such, a lot of studies focus on surrogate outcomes instead of the things that really matter to patients.

For example, in the course I teach on understanding medical studies, I interviewed a bunch of Yale undergrads asking them how important their grade point average was to them. You won't be surprised to hear that these young Ivy Leaguers, in general, considered the GPA to be a highly important metric—one that they were very concerned about. I next asked why it was so important to them. While some admitted there was a sense of personal achievement therein, the vast majority said that a good GPA would help them get a job they want or into that competitive graduate program. I then asked this follow-up: "Would you rather have a high GPA but not get your dream job, or a low GPA and get your dream job?" They all said they would rather have the dream job.

In this example, the GPA is a surrogate outcome, important only insofar as it is linked to something that is *actually* important:

getting a job. The GPA feels "close enough" to the actually important thing (the job) and is certainly easier to measure. (Students get GPA updates throughout their time at the college and get that first job only once.) Formally, a perfect surrogate outcome would be one that is tightly linked to the important outcome. Changes in the surrogate outcome should lead to changes in the important outcome. And perhaps most importantly, *anything* that changes the important outcome should also change the surrogate.

You see here why GPA is not a perfect surrogate outcome. Sure, people with higher GPAs tend to get better jobs. Sure, your GPA dropping can make your job prospects bleaker. But are there *other* mechanisms to getting a good job besides having a high GPA? Of course there are. You might be uniquely talented in computer programming, or a capable artist, or a passionate reformer, or the son of some well-connected politician. There are multiple paths to success, and not all flow through the GPA. I'd love to give you an example of a perfect surrogate outcome, but I can't. There really aren't any. But some are closer than others.

Yet a surprising amount of medical care is driven by surrogate outcomes. In fact, one of the main reasons doctors end up changing their minds is because they trusted a study that used a surrogate outcome, only to have a more definitive study, looking at a truly important outcome, overturn the prior finding. Both patients and doctors tend to put more stock in surrogate outcomes than we should.

For example, why do you want your blood-pressure numbers to be in the normal range? Do they mean something to you? Or do you want them to be normal because you want to lower your risk of heart attack or stroke? Ditto for cholesterol levels, vitamin levels, and body mass index. One could even argue that the size of a tumor is merely a surrogate for the risk of death from cancer. Would you rather die with a small tumor or live with a large one? One of my mentors early on told me there are only three outcomes that really matter in Medicine: birth, death, and quality of life. If we are not talking about one of those three things, we are probably talking about a surrogate of one of those three things. And that can be a real problem. Because surrogate outcomes don't always mean what we think they mean. In fact, sometimes using a surrogate outcome kills. I'll give you an example, using a drug that did exactly what we thought it would do.

Survivors of a major heart attack often die a short time later. This is an observation that has held from the time of Hippocrates, but it was only in the modern era that we began to understand why. After a major heart attack, patients are at risk of abnormal heart rhythms, some of which (like the dreaded ventricular fibrillation) are fatal. Arrhythmia is linked to death. Fortunately, in the mid-1980s, some new drugs for arrythmias were approved. They were approved on the basis of rigorous clinical studies that showed that they suppressed abnormal heart rhythms after heart attacks. What could be wrong with that? People who had more abnormal rhythms were more likely to die, and these drugs stopped abnormal rhythms. The minds of doctors were made up: These were good drugs that would certainly save lives. Prescriptions flowed.

But there was a problem: Arrhythmias are linked to death, but they aren't the same as death. They are a surrogate outcome for death. In 1989, a study was published in the *New England Journal* that looked at the effect of these anti-arrhythmia drugs on that truly important outcome: mortality. The 2,309 participants, who had recently had a major heart attack, were enrolled and randomly assigned an anti-arrhythmia drug or a placebo pill. Sure enough, the drugs stopped the arrhythmias, just like they were supposed to. But they also killed people. Five percent of people given the drugs died. One percent of people given a placebo died. Doctors' minds would have to change, and quickly, or more people would die.

Fortunately, they did. Prescription rates dropped precipitously, and subsequent research found medications (such as beta-blockers) that really do prevent death after a heart attack. Thank goodness we changed our minds.

This example illustrates two key points. First, it shows us that a study focusing on surrogate outcomes is not enough. Definitive studies examine definitive outcomes. Second, it demonstrates how important these types of trials are. A 5 percent death rate versus a 1 percent death rate is dramatic, but doctors were treating their patients with these drugs for *years* without raising concerns. And, honestly, we wouldn't expect them to. For any individual doctor, seeing maybe fifty or one hundred of these patients, a handful of deaths doesn't seem that crazy. After all, the patient had just had a major heart attack; it would be easy to rationalize a sudden death. It takes rigorous data collection to get to the truth.

Now, I don't want to give the impression that surrogate outcomes are *always* misleading. In fact, sometimes they work just fine. You can see a powerful example of the successful use of surrogate outcomes in the beginning of the HIV pandemic. The severity of HIV infection can be measured via two surrogates: viral load (how much virus is in a certain amount of blood; higher is worse) and CD4 count (the number of immune cells that are the target for the virus; lower is worse). When AZT, the first anti-HIV drug, came out, it was approved on the basis of the fact that it increased CD4 counts in infected individuals. Later, it would be shown to prolong their lives as well.

It's important that we not discard any study simply because it examined a surrogate outcome. Surrogate outcomes are often advantageous steps in the research process, because they provide an early read on whether a drug or intervention might be promising. The key is to understand that a study with a surrogate outcome is never the definitive study. In a perfect world, that definitive study would follow close on its heels. When it doesn't, it's up to all of us to demand that those studies be conducted. It doesn't mean we can't use the promising drug, but we need to use it with our eyes wide open to the possibility that we are improving a number, not necessarily a patient. And, importantly, when that definitive study arrives, we must all be willing to change our minds.

Three Words That Let You Know You've Got a Great Doctor

Most of the time, when doctors change their minds, it is for a good reason. We are trying our best to stay at the forefront of knowledge about health and disease, and that forefront is ever-changing. If we are being honest with ourselves and our patients, we can say the three most powerful words in Medicine: "I was wrong."

You want your doctor to be comfortable saying those words because all doctors will be wrong sometimes. We can be overtly wrong—ordering the wrong drug for the wrong patient or making the wrong diagnosis—and we must be completely transparent when those mistakes occur. But even the most conscientious doctor can't see into the future of medical science. We can't know what the best possible treatment will be ten years from now. All we can know for sure is that it will probably be different and better than what we are doing today, which is, in turn, different and better than what we were doing yesterday. Yes, we are always changing our minds. Yes, it may seem fickle. But in reality, it is growth. It is the evolution of Medicine into a more perfect science.

There was about a year of my practice in which I took all of my patients receiving one type of blood-pressure medication and switched them to another type of blood-pressure medication, as new data suggested the former wasn't as good at preventing heart attacks. Most of my patients didn't like this. They wanted to stick with what they knew—what they had been taking for years, in some cases. And that made sense. I needed to rely on the trust we had already built to say, essentially, "Listen, we've been through a lot together. I want you to give this change a shot because I think it will help you. If it doesn't, I'll be here for that too."

Many doctors feel that these types of changes are admitting defeat somehow, compromising our air of infallibility. Well, you know what? We are all fallible. The willingness to course-correct, if done collaboratively, honestly, and transparently, can actually *build* trust.

In sum, changing one's mind, whether as a patient or a doctor, can be one of the most powerful tools available to improve health, but it needs to happen for the right reasons. When we change our minds based on the incessant repetition we get from social media or even traditional media, we run the risk of succumbing to false beliefs. Instead, we must be like Bayes—evaluating new data as it comes in, and allowing that data to slowly, methodically shape our conclusions. When a doctor and patient can engage in this process collaboratively, with mutual trust, the path toward greater understanding is a veritable superhighway.

CHAPTER 3

The Temptation of the "One Simple Thing"

MR. AMBROSE WAS DYING. He had been dying for a while, hospitalized for about four weeks. I became involved in his care as his kidneys began to fail at the end of a long battle. It had started with pneumonia and was ending with multisystem organ failure—in other words, his body was shutting down, one organ after another.

There was no real hope for Mr. Ambrose. At ninety-two years old, in his condition he was simply too far gone to recover. I walked into the room to find him intubated and comatose. His blood pressure was being kept elevated with adrenaline and several other medications, but he made no response to the sound of my voice, to touch, or even to pain. He was dying—as many have died, and many more will—in a hospital connected to multiple machines. His skin was pale and swollen, like uncooked dough, his bodily fluids losing track of where they should be and collecting in his lungs, his abdomen, his legs.

I was slightly heartened to see a priest in the room, sitting with a woman in a wheelchair, who I presumed was Mr. Ambrose's wife. It seemed like they were coming to the end of their conversation, so I hung back, not wanting to interrupt the man who could probably do more good than I could in the moment.

"I just don't understand why it would have to happen like this. Why does God do this?" This from the elderly woman, whose stooped posture was likely the result of osteoporosis, but to me read as profound world-weariness.

"We can't always know the answers to those questions." This from the priest.

I admit I had hoped for something more compelling.

"I suppose you're right," the world-weary woman said.

"Take care. It looks like the doctor is here."

The priest made his exit, offering me a knowing smile. Doctors and clergy spend a lot of time in the same foxholes.

I sat down next to Ms. Ambrose, who said, sotto voce, "It's always the same shit with them."

I laughed out loud, which was probably inappropriate, given the circumstances. "You must be Catholic," I said. As a lapsed Catholic myself, my Catholic radar is strong.

"Oh yeah," she said.

"I hear you." I shook her hand, noting that her skin was dry and thin, like rice paper.

It had been a while since I'd thought about my Catholic upbringing. When I was eight or nine, in Sunday school, our teacher told us the story of the loaves and fishes. Lapsed Catholic version: A few thousand people have gathered around Jesus as he is preaching kindness and whatnot, but they don't have anything to eat. All the disciples can do is scrounge up a few loaves of bread and a couple of fish. Through a miracle, Jesus is able to feed all the thousands of people from this tiny supply. Our Sunday school teacher offered an interpretation. Perhaps, she said, this wasn't really a miracle of the flash-in-the-sky variety. Perhaps the miracle is that Jesus convinced everyone to share with one another, and that those who had more shared with those who had less, allowing everyone to eat.

Her interpretation really resonated with me. For the first time, the Bible didn't seem like an accounting of various magic tricks. To me, making bread appear out of thin air was not nearly as impressive as having the power to inspire others to kindness and compassion. This was a Jesus I could get behind. I didn't realize it at the time, but the story was the start of my understanding that hope can come in multiple varieties—the magical, miraculous kind, which I found (and still find) thin and unsatisfying, and this other more mundane type, which, to me at least, felt more genuine.

My Sunday school teacher was fired the next week. This line of teaching was definitely *not* Catholic orthodoxy. I think that's where my disillusionment with the Church started. I didn't go back to Sunday school after that. Since that time, Catholicism has not played a major role in my life, except in the occasional meeting with another lapsed Catholic like Ms. Ambrose. When two lapsed Catholics meet, there is a certain understanding that we share—a feeling that, yes, we've lost something along the way, but honestly what did they expect? They don't exactly make it easy to be a Catholic these days.

I introduced myself to Ms. Ambrose, who I found out quickly was Mr. Ambrose's sister, not his wife. A family of three siblings, two girls and a boy, none of whom ever married. The three of them had lived together in a small house in Hamden, Connecticut, for more than fifty years. They had no other relatives, no children, only each other. The older sister, I was told, had died last year.

And now Ms. Ambrose was sitting by her dying brother's bedside. She would be the last of the line. I sat longer with Ms. Ambrose than I normally would, bonding a bit over the nonsense of priestly celibacy. She told me about the Latin Mass, which I was too young to have experienced. "Pretty boring, really," she said.

I asked her if there was anything I could do for her. She asked for a cup of coffee, so I ran down to the cafeteria and got us each a cup. As we spoke, I became concerned about what would happen after her brother died. He had handled the finances, apparently, the shopping, the day-to-day. And when he was gone, there would be no one to help Ms. Ambrose. He died a few days later, with his sister beside him. I made a point of contacting the social worker and asking if there was a way someone could check on Ms. Ambrose in a week or two, and she said she'd let elder services know. I assumed I would not cross paths with Ms. Ambrose again.

But I did. A few months later, I was the attending physician on one of the inpatient medicine services and saw Ms. Ambrose's name on my list. Since my primary job is research, I spend only about eight weeks of the year caring for patients in the hospital—I'm not a fixture there, so it struck me as significant that our lives had intersected again.

She had been admitted to the hospital severely dehydrated and malnourished. Without her brother to take care of things like shopping and paying bills, she had languished. Since she was a wheelchair user even before she lost him and was now without regular support, I was actually surprised she had been able to live on her own as long as she did. The team took good care of her in the hospital, treated a brewing urinary tract infection, and started working on getting her nutritional house in order.

However, when I spoke to her, it was clear that there was something missing now. Not her brother, really. Or not precisely. But some motivating force, something inside us all that keeps moving us forward, keeps us eating, keeps us breathing. Ms. Ambrose told me she really didn't have anything to live for. She had lost hope.

We spoke often during her admission. Mostly, I was trying to convince her to go into an assisted living facility—an idea she loathed. The prospect of leaving the house she had lived in for decades, with all its memories, weighed heavily on her, but it was really the only option. Yet she was stubborn and savvy and difficult to persuade.

Until one day I mentioned that I'd visit her when she left the hospital. To bring hot dogs. She had confided that these were her favorite food, and I happened to live just minutes away from Blackie's, one of the all-time-great hot dog places in Connecticut (worth a stop if you're in the area). It was a little thing—the promise of a hot dog, or maybe the visit from someone who had become a friend—that changed her perspective and, I hope, brought a bit of brightness to a world that had felt meaningless since her brother died.

I took my kids to see Ms. Ambrose at the assisted living center two weeks after she was discharged, with enough hot dogs for all of us. Thankfully, she was thriving there. She had made friends. She was joking with the nursing staff. Her big, sardonic personality—the thing that had attracted me to her initially—worked on others as well. With a little bit of support, she was able to thrive again, and my impression was that despite the fact that she was living in a "managed care" setting, she felt freer than she had in quite some time. I knew she would never go back to living alone. But she was happy.

It turns out that sometimes you need something to hang on to—a dash of hope, a soupçon of meaning—to keep you moving forward one day at a time. That hope doesn't have to be a miracle, a flash in the sky, a cure, a restoration of youth. Sometimes that hope can be a hot dog and a chat with a friend.

Hope and Meaning in Medicine

One of the greatest challenges a doctor faces is to provide hope while still presenting an honest assessment of the clinical situation. Sometimes it is an impossible task. It's one thing to provide hope that your patient will get over their pneumonia. How do you provide hope when you need to tell someone that they have a terminal cancer? Some doctors believe that you simply have to find a way, pointing out that providing hope is not just kind; it is therapeutic. It's true that multiple studies have shown that cancer patients who remain hopeful, even with a terminal diagnosis, live longer, better lives. Yet providing hope sometimes means providing false hope, and that is something I've never been able to do. Maybe my patients would like me more if I did. But I wouldn't like myself very much.

I was moonlighting in the cardiac care unit one night when a twenty-five-year-old kid (at least, he looked like a kid) came in. He had a congenital heart condition and had been in and out of hospitals all his life, and at an accelerating pace in the last year. He needed a heart transplant, but there wasn't one available. When I went to examine him, he was scared. The first thing he said to me was, "I think I'm gonna die."

I see no need to build suspense here. I'm not sharing his story for its narrative arc. He did die—a few days later. But I'll never forget his fear that night. And more than the fear, the disconnect between the fear and what I was seeing in front of me. By the standards of the cardiac care unit, he was fairly stable. He needed a bit of oxygen, and his blood pressure was on the low side, but I didn't think he would die. And I told him that.

Yet my words and reassurance didn't help him. In retrospect, I assume this is because he knew he was sicker than I did, and my reassurance therefore seemed hollow, maybe even callous. My hope felt, to him, like false hope. So he remained afraid that night, lying alone in a hospital bed, while I made my rounds through the unit. There aren't too many medical experiences that keep me from sleeping at night, but this is one of them. His knowledge of his own mortality, his helplessness in the face of death, my inability to reassure him—all of these create in me a morass of guilt I feel to this day. I should have stayed in his room that night, talked with him about other things, distracted him, made jokes, watched TV. But I told him he'd be fine, as I often tell patients they'll be fine, and went on to the tasks I had set for the night.

I've told his story to many of my friends in Medicine, and almost all of them have had a similar experience. In fact, the "sense of impending doom" is taught in medical school as an early sign of a variety of life-threatening conditions, ranging from pulmonary embolism to cyanide poisoning. And, to be clear, not everyone who fears they may die actually dies—but it does happen. People do sometimes know. And at that moment, when death is staring us in the face, hope can be unattainable.

I have come to believe that we desire hope, but what we really need is meaning. Hopelessness arises in dire situations, and being hopeful is not always rational or possible. Meaninglessness, though, is death before death. Reminding someone of the meaning of their life, even if they can't be cured, can be transformative.

Hope and meaning. Two powerful motivating forces in Medicine and in life. And like with all powerful forces, there are those who would exploit them. In Medicine, the purveyors of false hope are protean. And they often come in the same form: the quick fix. The "one simple thing" you can do, or take, or change, to cure whatever ails you. Once you know to look for it, you'll see it everywhere. You can't miss it.

"One Thing" Medicine

I define "one thing" medicine as the idea that there is a single thing (a drug, a dietary intervention, a lifestyle change) that is guaranteed to change a specific health outcome. For a patient suffering from disease or disability, it is an incredibly attractive concept, because it links a simple change to a lasting benefit. The equation has sold millions of books and billions of pills. And, with rare exceptions, "one thing" medicine doesn't work. There are vanishingly few quick fixes in Medicine today.

Many patients are hoping for a miracle. And medical miracles do happen. But they don't happen through magic. They happen, more often than not, after decades of laborious research, of wrong turns, errors, and missteps. Medical miracles are hard-earned. It is for that reason that I find the idea that one simple thing can cure your ailment frustrating. Not only does it belittle the true miracles that are out there, but it overshadows them. The bright, shiny promise of the quick fix can lead people away from the real fix. We need to recognize the hype and beware of easy promises.

I should note, though, that the history of Medicine is littered with breakthroughs that might sound like quick fixes. Think of Ignaz Semmelweis convincing obstetricians to wash their hands—a practice that cut maternal mortality in half. Or think of the discovery that ether could put patients into a sleeplike state where they didn't feel pain, transforming surgery from distilled barbarism to the elegant art form it is today. Think of Alexander Fleming and penicillin, or Jonas Salk and the polio vaccine.

But as our understanding of human physiology and pathophysiology has developed, the low-hanging fruit has increasingly been plucked. We lead the long, relatively healthy lives we have now because of a century of medical discovery. We are the beneficiaries of hundreds of quick fixes that we now refer to as "standard of care." Breakthroughs like those still happen, of course. But "one thing" medicine isn't really about breakthroughs. It's about easy solutions to complex problems.

In Medicine and in life: If it seems too good to be true, it probably is. A wrist magnet cannot cure your arthritis. Increasing your quinoa intake won't prevent a heart attack. A multivitamin won't make you more energetic, and avoidance of red dye #40 will not make you immune to cancer. Those who offer easy solutions to difficult problems should be viewed skeptically.

It's easy to view other people's "one things" as ridiculous. It's harder to be honest about our own. I think the key to understanding this bit of psychology is to realize that "one thing" medicine is empowering. The one simple thing you do—your little ritual—makes you feel that you are in control of your health. I use the word "ritual" quite deliberately. In an era when, even in houses of worship, the ephemeral mystery of ritual is lost, people gravitate to small gestures of control over an increasingly complicated universe.

If you think hard enough, you will find your own "one thing," though you may not call it that. For me, it is doing push-ups in the morning and at night. I tell myself it's to keep my upper body strong, but, honestly, what are those push-ups really doing? They are giving me a moment to exert control in my life, to take charge of something that I know, rationally, would require a much greater investment of time and energy to really change. Those push-ups are my "one thing"—my irrational ritual that I do because it makes me feel better that I do it. There are all sorts of "one things" out there, but the area that dominates "one thing" medicine is clearly diet.

The One Diet/Supplement That Will Change Your Life Forever

Kale. Isoflavones. Red wine. Caffeine. Garlic. Each of these, and hundreds more dietary components, are aggressively marketed as essential to improving health and well-being. Why? Well, for the most cynical among us, it may be relevant to note that dietary supplements are not regulated by the US Food and Drug Administration. I could, right now, create a mix of various herbs and vitamins, blend them together, pop them into a pill, slap a fancy label on them that says something like MAXIMUM POTENTIAL, and sell them across the country. The key? I can't explicitly *say* that my new pill is intended to diagnose, prevent, treat, or cure any disease. But that's not really a problem. If I use a phrase like "supports your immune system" or "increases energy," have I suggested that my pill diagnoses, prevents, treats, or cures any disease? The FDA doesn't think so. And that's exactly why these phrases appear so prominently in the vitamin and supplement aisle.

This is not to say that all these pills and poultices are modern-day snake oil. Many of the claims made, whether on a vitamin bottle, or in the breathless reporting of a healthy-lifestyle blog, have a basis in the peer-reviewed medical literature. But these are almost always studies that show biologic plausibility, not true efficacy. The fact that an extract of kale gooses the antibody production from white blood cells in a test tube is certainly interesting, but it doesn't mean eating kale will improve your chances of fighting off the flu, or the common cold, or COVID-19. Beware biologic plausibility—remember how we have been burned before.

Not all vitamins, supplements, and dietary interventions exist merely to thwart FDA regulation. Plenty of studies of the effects of "one diet things" are conducted by well-meaning researchers who are trying to make our lives better. The problem is that many of them are using a really bad tool to do their work. The tool is a survey known as the "food frequency questionnaire."

How often do you eat bacon? Never? One to three times per month? Once a week? Two to four times per week? Once a day? Two to three times per day? Four to five times per day? Six or more times per day? That is just one example of the more than one hundred questions on a typical FFQ. If you were filling one out, you might be asked the same question about liver, fish cakes, sausages, cornflakes, brown rice, cottage cheese, bean sprouts, Ovaltine, wine, or basically any other food or drink you can think of.

The reason these surveys are so popular is because they are fairly easy for people to answer. I'm not asking you to remember whether you ate bacon three days ago. I'm asking, in general, how often you eat bacon. And most of us have a sense of that. FFQs aren't that accurate, though. When compared to a more reliable source (a food diary, kept in real time), food frequency questionnaires have, at best, modest accuracy. Still, they are the standard way that diet studies (nutritional epidemiology studies, to those in the know) are conducted.

And they are a major problem. The problem lies within all those different food types.

Imagine you give a food frequency questionnaire to ten thousand people. And then you follow that group for ten years. Lots of things are going to happen to them. Some will die. Some will have heart attacks. Some will get divorced. Some will get a new job. Some will disappear under mysterious circumstances. Each question—the bacon question, the bean-sprout question, the wine question—can be tied to any one of those outcomes in the form of a hypothesis. *I wonder*, you might think, *if bacon consumption reduces the risk of divorce.* Or maybe you could test whether bean-sprout consumption is associated with a reduced risk of heart attack. With a few hundred food items and a few dozen outcomes of interest, one large FFQ study can provide fodder for thousands of medical research papers.

And it doesn't stop there. You don't have to look just at the link between one type of food and a given outcome. Researchers have coded the food frequency questionnaire to allow the combination of foods to create new metrics—like total calorie intake, fat intake, carb intake—even *pesticide* intake. No, the FFQ does not ask how often you eat pesticides, but researchers know which fruits and vegetables tend to require more pesticides, and can thus use your FFQ answers to estimate how much pesticide you may consume. If this sounds crazy, because it doesn't account for things like washing your fruits and vegetables, that's because it is.

The more questions you ask, the more likely you are to get the wrong answer on one of them. Food frequency questionnaires give you the ability to ask a *lot* of questions. And, sure, most of the time you'll get the right answer. (There is no link between bacon and divorce rates, for those of you worried about this.) But ask enough questions, and you'll get the *wrong* answer. And when the wrong answer suggests that "one simple food" can improve your health... well, it's going to make headlines.

Take this example, which was presented at the prestigious American Heart Association national meeting in 2020. Researchers examined FFQ responses from around 570,000 people around the globe. They found that those who frequently ate chili peppers had a 25 percent lower risk of death from any cause compared to people who never eat chili peppers. Twenty-five percent! And all you have to do is suffer frequent taste-bud singes and the inevitable... ahem... secondary pain that occurs the next day.

Is this biologically plausible? Sure, spicy foods might decrease salt intake, for instance, and reducing salt intake is pretty good for you. But as we learned in the last chapter, biologic plausibility is a pretty low bar. Lots of things are plausible. It is also possible that people who eat a lot of spicy food do *other* healthy things—maybe they eat more vegetables, or live a more agrarian lifestyle, or live in countries with different or more equitable healthcare systems. The point is, the chance that the "one thing" is eating chili peppers is pretty low.

That, of course, doesn't stop the news media. Stories like this get clicks because we all want to know what the "one thing" is. That single food frequency questionnaire study generated many headlines:

EAT MORE CHILI PEPPERS AND YOU MIGHT LIVE LONGER—*Galveston County Daily News*

CHILI PEPPERS FOUND TO REDUCE CARDIOVASCULAR RELATED DEATH—*UK Diabetes*

LIVE LONG AND EAT CHILI PEPPERS?—*Forbes*

PEOPLE WHO EAT CHILI PEPPER MAY LIVE LONGER?—ScienceDaily

(Side note: I have a hypothesis that if you see a headline that ends in a question mark, you can just answer no in your head and skip the article 90 percent of the time.)

There are two possibilities to consider whenever you come across a "one simple thing" story. The first is that the study referenced therein is a false positive—a blip on the statistical radar. A coincidence, in other words, being amplified because of its clickability or, in darker cases, by individuals trying to profit off said "one thing."

There's another possibility, though. It may be that the "one thing" is true but simply doesn't matter.

The Signal and the Noise

Imagine you are sitting on the edge of a large pond. The day is deathly still. There is not a leaf rustling on any tree. The pond is as smooth and glassy as the surface of a mirror. A bird drops a seed into the middle of the pond, and ripples circle out from its point of impact. At the side of the pond, you see the ripple lap up at your toes, and infer (correctly) that those ripples are the result of the seed being dropped in the middle.

Now imagine the same pond on a stormy day. It's a big storm, and branches are breaking off trees. Wavelets are forming across the water as gusts of wind twist this way and that. A bird drops a seed into the middle of the pond, and you stare at your feet. Do you see the ripples? Can you tell which ripples come from the seed being dropped and which are from the chaotic turbulence of the storm around you?

This example is one I use to explain "signal-to-noise ratio" to my students. In both cases, the same event occurred (the seed being dropped by the bird). Against a background of very low noise, the signal of the seed being dropped (the ripple) is easy to detect. In a noisy environment (the storm), it is basically impossible to detect.

And yet the seed *does* affect the water. Something real *is* happening there. There is simply too much noise to detect it.

So how can you detect cause and effect in a noisy world? One way is through repetition. Create a system where the seed is dropped one thousand times, or ten thousand times, or more. Have a sensitive detector on the shoreline measuring ripples, and use mathematics to tell you if, on average, there is perhaps a slightly higher degree of rippling when you drop the seed compared to when you do not.

Some "one thing" studies are really using that process—aggregating mountains of individual-level data to detect a real, but frankly unimportant, effect.

While I do enjoy a littoral example from time to time, perhaps a more practical one would suit here. In 2015, a study was published in *JAMA Internal Medicine* examining the dietary patterns of almost one hundred thousand Seventh-Day Adventists. For those not familiar, Seventh-Day Adventists are a Protestant religious group that started in the United States in the latter half of the 1800s. They are characterized by their practice of worshiping on Saturdays, as well as their emphasis on a healthful lifestyle, which promotes vegetarianism and abstention from drugs and alcohol. Clean-living folks, basically.

The participants in the study each took a food frequency questionnaire, and researchers then followed them to see who would develop colorectal cancer in the next decade. They found that those who ate a predominantly vegetarian diet were less likely, on average, to develop colorectal cancer than those who ate more animal products.

This study was far from perfect. People who eat vegetarian (particularly when it is a tenet of their religion) are more likely to engage in other healthful behaviors than those who do not. But for the sake of argument, let's say that the underlying message of the study (eat vegetarian to reduce your risk of colon cancer) is true.

Should this be your "one thing"? If so, there's a problem. The cases of colorectal cancer were lower in the vegetarian group, but they were extremely low overall. Of the nearly one hundred thousand people in the study, 0.62 percent of vegetarians developed colon cancer (over seven years of follow-up) compared to 0.64 percent of the nonvegetarians. That means that, if this effect were real, you'd need to convert about five thousand people to vegetarianism to prevent one extra case of colorectal cancer. In other words, the chance that this "one thing" would be the one thing to save *your* life from colon cancer is vanishingly small.

Interestingly, the rates of colorectal cancer in the Seventh-Day Adventist population are substantially lower than in the general US population. You'd need to convert only three thousand people to Seventh-Day Adventism to prevent one case of colorectal cancer, two thousand fewer people than you'd need to convert to vegetarianism to get the same result. Praise be.

Here's the truth: There is no "one thing" that will help you live longer. There are healthy things, and less healthy things, and unhealthy things. People who adopt multiple, broadly healthy lifestyle choices—from what they eat and don't eat, to how much exercise they get, to the kinds of activities they take part in—live longer. The problem is no one wants to click on an article saying MAKE THESE 30 CHANGES IN YOUR LIFE TO LIVE LONGER. It is not easy to make thirty lifestyle changes. We all want simple solutions to complex problems.

I should note that "one things" are not limited to vitamins and minerals and health food supplements. Pharmaceutical companies absolutely want you to believe that the drug they are marketing is the "one thing"—at least for *your* disease. In 2010, an ad for the cholesterol-lowering drug Lipitor appeared in *Time* magazine. The prominent text read "A lot of people think exercise and healthy diet are enough to lower high cholesterol. For 2 out of 3, it may not be." In other words, don't worry about that other stuff; this "one thing"—our drug—is the real key to getting your cholesterol under control.

I think that deep down, we all know that a simple solution will rarely solve a complex problem. So why do we continue to adhere to these "one thing" beliefs? How do they enjoy a special place in our otherwise rational brains? If we believe eating chili peppers will make us 25 percent less likely to die, how do we explain that everyone else isn't doing it? Why are there no chili-pepper-recipe public service announcements? Why doesn't Medicare send chili peppers to all of its beneficiaries?

To continue believing in our "one thing," we need to surround and defend it inside an infrastructure of supporting beliefs. Perhaps the "one thing" works, but the knowledge is being suppressed by shady, ambiguous forces. Or the "one thing" works, but only if used in a certain way. In other words, if you imbue the "one thing" with an aura of secret knowledge, it becomes much easier to believe in it.

The Allure of Secret Knowledge

A "one thing" philosophy gives you a sense of control over things that are difficult to control, but there is more than just a desire for control behind people's eagerness to embrace the "one thing." There is the allure of secret knowledge: the appreciation of a deep, fundamental truth known only to a small group of individuals. When anyone suggests that they have a simple "one thing" solution to a complex medical problem, I am immediately skeptical. Not because I think they are grifters (though some are), but because I know how easily that way of thinking can become entrenched. Being "in on the secret" can become an identity in and of itself. You start to feel the truth of the "one thing" deeply, in your bones, and are often quite loath to accept explanations that counter the underlying premise.

No doubt the strangest "one thing" I came across in my medical career presented itself as a medical mystery. A woman was referred to our kidney clinic because her blood work was consistent with kidney failure. To my surprise, when I entered the room, I found a well-appearing older Russian woman with no specific complaints. She said her appetite was good, she was urinating normally, and she had good energy. But when I examined her blood work, I found her creatinine level to be extremely elevated. For context, a normal creatinine level is about 1. Above 1.5 is consistent with kidney disease. Above 4, we start talking about dialysis or kidney transplant. Her level was 10.

Yet the rest of her labs looked pretty normal. People with advanced kidney disease tend to have low red blood cell counts, but hers were just fine. Advanced kidney disease leads to a buildup of urea in the blood, but hers was normal. No, it seemed clear that the creatinine level was an aberration. But why? How does a healthy person get their creatinine level to 10?

It turns out it was her "one thing": ultracondensed beef broth, which she believed kept her strong, kept her immune from infectious diseases, and would allow her to live (as her parents did) well into her nineties.

Creatinine is produced in a relatively constant amount by muscle cells, filtered out of the blood, and then excreted by the kidney. Meat is muscle (though it isn't terribly appetizing to think of it that way), and so it also contains creatinine that must be excreted by the kidney. Normal ingestion of meat adds a trivial amount of creatinine compared to the ongoing activities of every muscle in your body. My patient, though, had found a way to ingest truly Herculean amounts of creatinine. She would boil beef broth down to a thick sludge, which she would then eat. The sludge was chockablock with creatinine, more than her kidneys could keep up with, so the creatinine level in her blood rose.

Fortunately, creatinine itself isn't toxic. It's just a useful marker of how the kidneys are doing. Her "one thing" didn't cause her any great harm, aside from a trip to the kidney specialists' office. But what stuck with me was her passion about this "one thing." To her, the condensed beef broth was the secret to a long life and to boundless energy. There was no way I could convince her otherwise, and, provided she knew how to explain her weird kidney numbers to other doctors, I wasn't particularly inclined to try.

In some sense, the "one thing" was part of her identity; it was something that felt deliberate and meaningful to her. There is a sense of ritual there too—the painstaking and meditative process of reducing beef broth giving her a dedicated time (once a week, she told me) to consider her future. We should all be so lucky to have that kind of a reflective carve-out.

Why "One Things" Feel Meaningful

The allure of secret knowledge is more than spiritual; it is biochemical. In fact, our brains are wired to search for these deeper truths. And this wiring can be exploited. To understand how this works, you need to know that your brain is constantly forming memories. Attached to these memories are all sorts of tags and flags—metadata—that get stored with them, particularly your emotional state when the event in the memory occurred. But there is another bit of metadata that gets stored along with the memory: its meaningfulness.

Meaningful memories, naturally, are stickier than meaningless ones. I remember my wedding day vividly, but not the day I turned thirty-five and a half. This is a good thing. Our brains would get too full of meaningless memories if we remembered everything.

Meaning is generally attached to a memory based on the emotions experienced while the memory was formed. Heightened emotions—fear, excitement, anger—lead to more meaningful memories. But relatively recently, and through an unlikely source, the biology of meaning has been more deeply elucidated.

It turns out that there is a certain receptor in your brain, known as the serotonin 2A receptor, that when activated, adds the "meaningful" tag to memories. It can be activated through emotion, certainly. But it can also be activated by certain agents that bind to the receptor. The most well-known? Lysergic acid diethylamide, or LSD.

In 2017, researchers from Switzerland took twenty-two volunteers and had them listen to a playlist of songs. The playlists were made to contain some songs that were very meaningful to each volunteer, and some that weren't meaningful at all. Under normal conditions, the participants rated the meaningful songs as meaningful and the nonmeaningful songs as nonmeaningful—pretty straightforward. But then they were given relatively high doses of LSD. Now the outcome changed. The nonmeaningful songs became highly meaningful to the participants. The meaningful ones stayed meaningful too. In fact, everything was meaningful, from the music to the walls of the MRI machine, to the crinkles in the subjects' pants.

When mundane things—a song you don't particularly care for, or the cracks on the wall, or the pattern in your carpet—have *meaning* attached to them, it can feel otherworldly. You may feel you are in the presence of some greater power, perhaps God, for how else can you explain how looking at a crumpled piece of paper can seem so *important*?

It quickly became clear that this "meaningful" tag was added by activating the serotonin 2A receptor. When the researchers gave a drug that blocked the receptor at the same time as they gave LSD to the participants, the meaningfulness went away.

In fact, the inappropriate assignation of meaning to meaningless events may underlie some of the paranoia and delusions seen in patients with schizophrenia. Imagine if your "This is really important" sense was turned up to 11 all the time, no matter what you were doing. Might you not think you were the center of some vast conspiracy? That you were being followed? That messages were being sent directly to you? It is no wonder, then, that many of the antipsychotic drugs used to treat schizophrenia block that serotonin 2A receptor.

You don't need LSD to activate this receptor. Highly emotional, spiritual, religious, and mystical experiences appear to activate it as well. This serotonin-based system in the brain leads to a spectrum of phenomena—from simple memory (don't eat that; you got sick last time) to a feeling that a piece of information is important, to transcendental religious experience, and, in the extreme, to psychosis.

When you become aware that the sensation of meaningfulness can be mediated by a chemical process, you can begin to examine more critically those things in your life that you feel are meaningful. We are designed to believe ideas that come with the "meaningful" tag. Evolutionarily, this is critical. If our friend died shortly after eating some algae he collected at the seaside, it behooves us to find that memory meaningful and not go try the algae for ourselves.

But that feeling of meaningfulness can trap us in bad habits or false beliefs that preclude us from making the right choices about our health. The "meaningful" tag can be attached to drugs, to objects, to practices, to dietary elements, even to people. Charismatic individuals can leverage the emotional pathway to the serotonin 2A receptor to convince others that bizarre beliefs are true—cults like Jonestown and Heaven's Gate come to mind. But more mundane beliefs can become entrenched as well due to a sense of meaningfulness.

Talk to any cancer doctor, and you'll hear a version of a story that I heard from my wife a few years ago. A young woman was informed she had a new diagnosis of breast cancer. It was immensely disturbing to her—she felt like her world was turning upside down. The hopes and dreams she'd had for her future were crumbling before her eyes. She visited a medical oncologist who laid out the path of treatment: surgery, followed by chemotherapy, which would be accompanied by side effects including hair loss, nausea, and nerve damage. The chances of success were good, but not perfect. And even if the cancer were to go into remission, there would be a lifetime of monitoring for recurrence. She was scheduled to see my wife for the preoperative visit, and she didn't show up. Multiple phone calls went unanswered. Eventually, they gave up trying to reach her.

About a year later, she came back to the practice, now with metastatic breast cancer. Surgery was no longer an option. She had spent the year on a no-sugar diet—a "one thing" that felt right to her (and that is supported by an entire online ecosystem of promoters but scant scientific evidence). In order to believe that cutting all sugar out of her diet would cure her cancer, she'd had to add an ecosystem of ancillary beliefs, including the idea that doctors know full well about the healing power of this diet but don't tell their patients due to the need to keep money flowing into their pockets.

It isn't true, of course. As a researcher myself, I can tell you that all of us want to become famous by stamping out a disease and/or saving the world, and if I had secret knowledge of how to cure cancer, I would have blabbed about it long ago. But the need for "one thing"—one thing that would get her life back to normal, the way she'd imagined it would be—was simply too compelling to ignore.

No one deserves to be sick. So when we become sick, we feel that in a just universe, there should be an easy way to return to health. For what crime are we being punished? And if there is no crime, shouldn't we be set free? The lack of cosmic justice is something we see all too often in Medicine. In fact, the absence of karma is something of a dark joke in medical circles. After seeing a patient in the hospital for an initial consult, I'll debrief with my trainee in the hall outside or in the charting room. For some patients, the first comment is a worried "He was really nice." Left unsaid is something physicians understand within weeks of starting our jobs: Nice is bad. Bad things happen to nice people.

Not really. Bad things happen to everyone, regardless of how nice or not nice the patient is. But when bad things happen to nice patients, the memory gets the "meaningful" tag (due to the emotion associated with it), meaning we remember it more. It leads to this perception among doctors that nice patients get into trouble, and the shared worried glances when we leave the room of a particularly affable individual. There is no cosmic justice. Only people doing their best. Sometimes the hand you are dealt is a bad hand, but you still have to play with those cards.

Once you recognize the allure of the quick fix, whether it is in an ad or a Facebook post or a magazine article, you will be able to avoid it much more easily. You'll recognize those who promote quick fixes as (in the worst case) charlatans or (in the best case) people caught up in those incredibly powerful forces—the allure of secret knowledge, the power of meaningfulness—that can so easily entrap us if we are not on the lookout.

The downside is you'll realize that the real solutions are rarely simple. For some, that may be a bleak conclusion; we don't always have all the answers, and even when we do have solutions, real change requires real work. But for most of us, the realization can be liberating. It means we don't have to pore over the latest wellness blogs or articles, trying to figure out what the secret is. We don't need to throw away all the food in our cabinets and replace it with the latest "superfood." And, critically, it means we can speak with our doctors and face the harder truth of the treatments we have in front of us. We can trust them when they tell us that it won't be easy, that we have work to do, that there isn't always a straight path or one simple trick.

That's the secret I'm sharing with you now: There is no secret.

CHAPTER 4

The Quest for Causality

SOMETIMES, THE LINE between life and death is made of plastic. During my residency training, I spent weekend nights moonlighting in a three-hundred-bed community hospital about an hour west of Philadelphia. It was a typical nonacademic hospital, where private practice physicians would care for patients, and the day-to-day management fell to the nursing staff, who were given broad leeway. Private practice docs cannot spend all their time in the hospital, and, as such, except for one doctor in the emergency room, there were no physicians in the hospital from seven in the evening to seven in the morning. The exception was me—the moonlighter. My job was to be there in case, as the administrator of the hospital described it to me during my interview, "shit went down."

It was a pretty nice job. I'd arrive a bit before seven and tuck myself into a comfy call room on the ground floor, usually with a small supply of snacks I'd picked up at the service station on the way. I'd flip on the TV in the corner, maybe work on my laptop a bit, and, eventually, go to sleep. Until, at some point in the night, I would get a call. Something, somewhere in the hospital, was wrong.

While most of the emergencies were routine, every once in a while a truly frightening situation would develop. During those times, being the only physician in a three-hundred-bed hospital starts to feel very lonely. It was around one in the morning when the nurses woke me to check on Ms. Murphy. At eighty-one years old, Ms. Murphy had been in and out of the hospital for years due to emphysema and other complications from a lifetime of smoking. She had been admitted from the ER just twelve hours earlier with difficulty breathing, but some nebulizer treatments had improved the situation. Everyone had expected she would have a quiet night. But when her routine vital signs were checked, the nurses were shocked to see that the amount of oxygen in her blood—the "O_2 sat"—was frighteningly low, at 65 percent. (Normal is above 94 percent.) They had, appropriately, placed an oxygen mask over her nose and mouth and turned the wall oxygen up to twelve liters per minute—the highest flow it could achieve. But the oxygen saturation didn't budge. Ms. Murphy was in trouble. Anticipating that she would need to be intubated and placed on a breathing machine, they called me.

I ran to the bedside to find five nurses in the room next to the bright-red "crash cart," the ominous toolbox containing the medications, tubes, lines, and other paraphernalia we would need if Ms. Murphy stopped breathing or her heart stopped beating. She was awake but delirious and breathing rapidly, her lips and nail beds dusky from the lack of oxygen. I took stock of the situation as quickly as I could: Her heart rate was elevated, but her blood pressure was stable; that was a good sign. But I had the nurses place an emergency call to anesthesia for a potential rapid-sequence intubation.

In the meantime, I listened to Ms. Murphy's lungs. There was some wheezing there, but it wasn't as bad as I'd expected. Even individuals with severe emphysema should have a normal oxygen level when they are receiving twelve liters of oxygen through a face mask—why was hers so low? Importantly, I could hear air going into both lungs. Sometimes patients with severe emphysema can rupture a thin, scarred segment of lung tissue, leading to the collapse of an entire lung, but that was clearly not the case here.

It must be a pulmonary embolism, I thought—a blood clot that had traveled from her leg and lodged in her pulmonary artery, preventing blood from flowing easily through the lungs. *The air was getting in, but the blood wasn't.* If this was the case, it was bad news. Though there are treatments for acute pulmonary embolism, one that was bad enough to bring an O_2 sat down to 65 percent even on twelve liters of oxygen would prove difficult to survive.

I was about to start blood-thinning therapy when it got really quiet. You know when you have a bunch of people all in a room talking and, for whatever reason, all the conversations stop at the same time? That happened—just for a second. And the only sounds in the room were the beeping of the cardiac monitor and the whoosh of the oxygen. And I'm not sure why, but that whoosh did not sound quite right to me.

It was enough to make me sit back for a second and look carefully at the bed—a tangle of wires and tubes. I walked over to the wall to the oxygen valve—set to twelve liters a minute, flowing into a green plastic oxygen tube. I took hold of the tube and traced it, carefully, toward Ms. Murphy.

I found the end of the green tubing, plugged into a nasal oxygen loop, lying underneath the bed. Tracing the tube from the face mask Ms. Murphy was wearing, I found *its* end (the one that should have been plugged into the wall oxygen) also under the bed. There was no connection between the face mask and the wall oxygen at all. She was not receiving twelve liters of oxygen; she was receiving… well, just air.

As soon as I plugged the face-mask tube into the wall, the numbers started to improve. She perked up immediately, restored by the high oxygen flow. Her lips and nails brightened. Her heart rate slowed down to normal. Over the next hour, we were able to wean the oxygen down to just four liters. We ran confirmatory tests, of course (I was still worried about pulmonary embolism), but in the end the treatment she needed was the one we *thought* we had been giving her all along.

Ms. Murphy's case would spawn a lengthy safety review at that small hospital, and new training practices designed to ensure that similar mistakes didn't happen again. But what struck me about the case was how, for a moment in time, Ms. Murphy's life depended entirely on her connection to an oxygen delivery system represented by a small valve in the wall. So long as that connection was intact, she would live. But if that connection were severed, she would die. Scientists consider cause and effect to be connected, but usually not in the literal sense of the word. Here was a case where cause and effect were literally linked, via a thin green plastic tube.

The essence of Medicine is cause and effect. Practicing Medicine is a series of choices, each of which represents a moment in time—a cause—and each of which is linked to effects in the future. Understanding those links is what allows us to help our patients make good choices about their health. The problem is that the links are rarely as clear as a green plastic tube leading from wall oxygen to a face mask. In fact, many of the cause-and-effect links in health are opaque, mysterious, and difficult to measure. In other cases, we may believe that a given cause is linked to a given effect when, in fact, it is not.

Humans Are Causal-Inference Machines

The sun rises, and the rooster crows. You flip the switch, and the light turns on. You wear your lucky socks and land that job you were angling for. Humans are exquisite in our ability to connect A to B, even when no connection truly exists. But it is vital for us to attempt to make those causal connections, because, in general, they reinforce behavior that leads to good outcomes. At least, that was the case for the majority of human history. If our ancestors were foraging and came across a new berry—red, sweet, and delicious-looking—but became violently ill after eating it, they would not eat it again. Their brains appropriately concluded that the berry caused the illness. In Medicine, we often refer to this paradigm as the "exposure-outcome link." The berry is the exposure. The violent gastrointestinal illness is the outcome. Understanding cause and effect saved our ancestors' lives.

Of course, our ancestors made their share of causality mistakes as well. Though it is unclear what led to the start of human sacrifice rituals in pre-Columbian Mesoamerica, we do know that a major role for human sacrifice was to ensure that the gods were strong enough to stave off a great catastrophe in repeating fifty-two-year cycles. On the eve of such a cycle, all fires were extinguished in Aztec lands, and a human sacrifice was made. If the sun rose again the following morning, the gods were appeased, fires were relit, and a new fifty-two-year cycle could begin. In the minds of those ancient Aztecs, there was a causal link between human sacrifice and the world continuing to spin on its axis.

That belief was self-reinforcing. After all, the world *did* continue to spin on its axis. No one—at least no one that we have records of—suggested that the Aztecs skip the ceremony once just to check their assumptions. If they had, quite a few lives would have been saved. I often tell this story to students in the context of one of my favorite causality maxims: "We cannot assess causation without variation." In other words, we can't tell if A causes B if we never change A.

Our human tendency to infer causality is a double-edged sword. Coupled with our tendency for motivated reasoning, as discussed in chapter 1, cycles of false belief can be created. We want control over a chaotic world, so we decide to carry a lucky rabbit's foot or hold our breath when we pass a cemetery, falsely linking these actions with outcomes we view as favorable and ignoring or overlooking times when our expectations aren't met, which reinforces the behavior. We call these cycles of false belief "superstitions" if they do not significantly impact our day-to-day functioning, and "delusions" if they do. While minor superstitions, like knocking on wood, can be perfectly harmless, the false assignation of causality can sometimes lead to significant, even life-threatening problems.

I am reminded of one of my patients, who became convinced that his kidney disease was due to using sugar substitutes. While I didn't begrudge him his decision to use only natural sugar in the future, he refused to take medicines that could preserve his kidney function, assuming that the kidneys would recover once the exposure to Splenda and Sweet'N Low was removed. Unfortunately, they did not.

Medicine has gotten it wrong before. For over two thousand years, the most common medical procedure performed by surgeons was bloodletting—the draining of blood from the body of a sick individual in order to balance the underlying "bodily humors." With rare exceptions, bloodletting was harmful to the patient, and yet the practice continued, supported by tradition, "expertise," and humans' ability to ignore data that doesn't fit their underlying assumptions. If the patient recovered after a good bloodletting, it was further evidence for the utility of bloodletting. If they did not... well, they certainly would have died anyway. It is quite possible that George Washington died of this medical malpractice, as his physicians removed 3.75 liters of blood from his body to cure a throat infection. This represents around 75 percent of the total blood volume of a typical adult male. He expired shortly after the procedure, on Saturday, December 14, 1799.

When a patient asks for a treatment that is not causally linked to the outcome we are trying to achieve, a physician is obligated to refuse. We cannot ethically perform a bloodletting even if the patient asks us to. But is there anything more frustrating than asking your doctor to prescribe a medication and having them say no? Believe me, it is frustrating for the doctor as well. We don't like to say no to patients—it's not in our makeup—but sometimes we have to. Like most docs, I've refused patient requests for antibiotics for viral infections, as antibiotics do not have an effect on viruses. I've refused more esoteric requests, like the patient who wanted me to perform a spinal tap to see if she had Lyme disease in the brain, and the patient who wanted me to prescribe growth hormone to extend his life.

Doctors refuse treatments for two main reasons. The first is that the risk is too high (we all took an oath to "do no harm"). The second is that we don't think the treatment will lead to the benefit that the patient thinks it will. In other words, we don't think the treatment is causally linked to the outcome. Causality is everything to a physician; it explains so much of our reasoning that can otherwise seem strange or even irrational to a patient. As such, it is critical that patients understand how doctors think about causality, how we test it, and how we can sometimes get it wrong.

Recognizing Causality

Philosophers from Aristotle to David Hume have struggled with the concept of causality. You could spend semesters in college and grad school trying to grasp the formal logical links between cause and effect, and how they can be manipulated. Fortunately, this book is not a textbook and is not meant to replace one. For the purposes of Medicine and health, a simple definition of "causality" will suffice: If A causes B, then changing A will change B. If smoking causes lung cancer, then not smoking will prevent lung cancer.

You'll note that this is not precisely the definition of "causality" we use in everyday life. In our minds, causality emerges more as a feeling than a logical sequence of thoughts. In everyday life, we use several rules of thumb to determine whether an event (or outcome) was caused by some other event (or exposure). The most primal, visceral, and straightforward way humans infer causality is through a temporal relationship. I flip the light switch, and the light comes on more or less immediately. I infer that the flipping of the light switch *caused* the light to come on.

Now, the electricians among you will no doubt be tempted to correct me here. The light switch doesn't power the light but, rather, it allows electrons to flow and excite a filament (or LED), which generates the light that hits my eye, and so on and so forth. But those things—the electricity flowing, the heated filament—all lie on what we call the "causal pathway" from switch to illumination. The light switch might not be the most *proximate* cause of the light turning on, but it was the cause. If I *don't* flip the switch, the light doesn't turn on, after all.

In many ways it is easier to use temporality to *disprove* causality than the other way around. If I become ill on a Wednesday and eat sushi the following Tuesday, it is not possible that the sushi caused my illness, and our brains make no such association. But temporality can be highly misleading as well. I remember a story I heard once from Paul Offit, the noted pediatrician and vaccine researcher. He was seeing kids in his office at Children's Hospital of Philadelphia, and a baby boy, eighteen months, was coming in for a visit. He was scheduled to receive his first dose of the measles, mumps, and rubella vaccine. The little boy suffered the first seizure of his life while in the waiting room *before* the visit. He was successfully treated. But can you imagine if that first seizure had happened just ten minutes later, after the boy had received that first MMR vaccine dose? Even the most levelheaded among us would have a strong sense that the vaccine, in some way, *caused* the seizure. Yet the temporal relationship doesn't prove causality. It only supports causality.

Humans are also spectacularly bad at inferring causality when the outcome happens long after the exposure. There is no particular reason that causal events need to be linked by short periods of time. Some asteroid colliding with another asteroid millions of years ago might send one of the rocks hurtling toward Earth a million years from now, and those events are just as causally linked as the light switch and the lightbulb. But for humans, time is very important. We can lose motivation quickly if we don't perceive short-term gains from our healthful behavior, and we may not concern ourselves with unhealthy behaviors like smoking if the outcomes won't occur for years or decades. We are much more likely to cut back on drinking alcohol to avoid suffering a hangover the following morning than to avoid liver damage that may not manifest for thirty years.

This phenomenon more or less explains our behavior when we are in our teens and twenties. It is why we have difficulty sticking with an exercise routine, or changing our diet, or studying for the SAT—the benefits of these actions are not immediately evident. It is also the reason we are so much more comfortable taking medications that address immediate symptoms (think ibuprofen for that headache or, more extreme, oxycodone for that back pain) compared to medications that are needed for some longer-term benefit (like cholesterol drugs to prevent a distant heart attack).

My heuristic from the start of the chapter (A causes B if changing A changes B) is good but not always practical to test—particularly when outcomes are very distant in time from exposures or when A can't be easily changed. For example, it is likely that healthy eating in your twenties is causally linked to better cardiovascular health in your seventies, but no one is going to fund a fifty-year trial, randomly assigning twenty-year-olds to various diets to figure out if they'll be healthier in their seventies. Or consider the causal link between growing up in poverty and future cancer risk: It is not as easy to change someone's economic status as it is to tell them to take a new pill. When possible, the best strategy to determine causality *is* to change the exposure and observe whether the outcome changes; this is fundamentally what happens when researchers conduct randomized controlled trials. But when that isn't feasible, we need some other tools.

In 1965, British epidemiologist Sir Austin Bradford Hill listed nine criteria he considered as evidence supporting causality. Temporality was just one of them. The "Bradford Hill criteria" have come under fire over the years but still remain one of the standards by which researchers explore causal hypotheses. I've always found it difficult to teach Hill's criteria, because there are many and there are few examples of hypotheses that satisfy all nine of them, but I'll give it my best shot.

The hypothesis of interest: Drinking alcohol at night causes a worse night's sleep. True or false? Causal or not?

Hill would analyze this hypothesis across his nine criteria, as follows.

- **The strength of the association**. Hill felt that if A causes B, the link should be strong. Do I have a restless night *every* time I have a glass of wine after 9 p.m.? Or only one out of ten times? The former makes a causal link more likely.
- **The consistency of effect.** If there is a causal link between alcohol and restless sleeping, it should occur not just in me but in other people around the world, and when investigated by other researchers.
- **Specificity.** This means that the outcome (restless sleeping) should very rarely happen when alcohol is *not* involved (a harder box to check in my case, given my tendency to be woken at 1 a.m. by small children who need a glass of water).

- **Temporality.** We know about this one. We'd want the alcohol drinking on Tuesday to affect the sleep quality Tuesday night, not Thursday or next year.
- **Dose response.** More alcohol should lead to worse sleep if the causal link is there.
- **Plausibility.** Another one we are familiar with. It is, of course, biologically plausible that alcohol could worsen sleep. (It can suppress REM, lead to dehydration, increase the need to get up to pee, you name it.)
- **Laboratory evidence.** Hill would want us to show that, in a petri dish, alcohol might affect neurons, for example.
- **Experiment.** I could randomly pick nights to drink and nights to abstain, and see how the sleeping goes.
- **Analogy.** This relationship is similar to another causal relationship (like the effect of drug use on sleep quality, for instance).

The Bradford Hill criteria are a steep hill to climb on our quest for causality, and I am not arguing that you need to apply the criteria whenever you are assessing whether there is a causal link between an action you take and the outcome you observe. But I find them useful to remind me that it isn't *just* temporality that matters. You should also note that just because an exposure-outcome pair satisfies *one* of the Hill criteria, it does not prove causality. A study that shows a dose-response effect between eating brussels sprouts and lower rates of gastric cancer might be interesting, but we would want to see more criteria met before we conclude that we should force the bitter veggies down our throats. Hill's criteria move the needle to a conclusion of causality, but no one criterion is perfect, and counterexamples to each of them are easily found.

For example, in terms of the strength of association, you will find few links more statistically significant than that between the consumption of margarine and the divorce rate in Maine (99.2 percent correlation, per Tyler Vigen's excellent Spurious Correlations website). And yet the strength of association alone is not enough to conclude that margarine is a major driver of divorce in the Pine Tree State.

Spurious correlations can be a major source of medical misinformation and false medical beliefs. So how can you extricate mere correlation from true causation?

Correlation Is Not Causation, But...

You've no doubt heard the old adage that correlation is not causation. Spurious correlations abound, as the Maine-margarine example shows. Vigen's website also notes the remarkable correlation between swimming-pool drownings and the number of films Nicolas Cage has appeared in, the number of murders via hot objects and the age of Miss America, and the number of engineering doctorates and the volume of mozzarella cheese consumption. (Come to think of it, that last one *might* be causal.)

But totally spurious correlations—random quirks and coincidences in data sets—are not likely to trip you up. The most common reason to falsely assume that a correlation is causal is because you forgot about something—a third factor that links the exposure and the outcome of interest. Formally, these are known as "confounders."

Why is the murder rate in a city correlated with ice cream sales? Is there a causal link here? Are murders committed by people driven insane by ice cream headaches, or does the sugar rush prompt impulsive behavior? Of course not. Ice cream sales are correlated with the murder rate because both things happen more in the summer (ice cream sales because the temperature is hot and ice cream is cold and delicious, murders because more people are outside, leading to more confrontations).

We say that outdoor temperature *confounds* the observed association between ice cream sales and murder rates. A confounding variable is one that is linked to the exposure of interest *and* the outcome of interest, and thus induces the *illusion* of causation between the exposure and the outcome.

Another example: In 1999, an article appeared in *Nature* (one of the most prestigious scientific publications) which found that children who slept with a light on were much more likely to be nearsighted when they grew up. The press latched on to this with a *causal* explanation that perhaps the light exposure at night was limiting eye growth in a way that would promote poor vision. In fact, there was no causal relationship between leaving the light on and poor vision. Future studies found that there was a confounding variable—parental nearsightedness—that wasn't accounted for. It turns out that parents who are nearsighted are more likely to have kids who are nearsighted. Parents who are nearsighted are also more likely to leave the lights on at night—the easier to check on the kids. In other words, there was a missing variable in the *Nature* analysis. If you account for parental nearsightedness, the implied causation between leaving a light on and child nearsightedness disappears in a puff of logic.

These third factors occur all the time in observational data sets—in fact, there is no way to fully account for all of them. There is a strong correlation between alcohol intake and lung cancer, but not because alcohol causes lung cancer. Rather, people who smoke are also more likely to drink, and also more likely to get lung cancer. Smoking *confounds* the observed alcohol–lung cancer relationship. If A doesn't truly cause B, changing A won't change B. Turning off the night-light won't stop kids from being nearsighted any more than reducing your alcohol intake will ward off lung cancer or eliminating ice cream sales will decrease the murder rate. Correlations can be interesting, but without causality they are not actionable.

Confounding will be present in any study that does not use randomization to assign the exposure of interest, and the vast majority of medical studies are *not* randomized. Therefore, the vast majority of medical studies assess correlation, not causation. They are, under the hood, an observation of data points that are examined mathematically for interesting relationships. Some of those correlations will be causal. Some will not.

Correlational studies often gain traction because they *feel* right to us. (That's motivated reasoning again.) In 2017, a study appeared in *Pediatrics* that found that babies who were breastfed longer had fewer "problem behaviors," higher IQs, and greater vocabulary when they grew up. Is it biologically plausible that more breastfeeding helps kids' brain development? Sure, there are no doubt lots of good brain chemicals induced by all that close contact with Mom. But when the authors of that study used statistical tools to control for confounders—like parental income—the correlational findings vanished.

Breastfeeding longer doesn't seem to *cause* smarter babies. Rather, breastfeeding longer is a marker of higher income, and with higher income comes a host of advantages to children, from access to better healthcare to private tutoring, and so on. Since correlation isn't causation, *changing* your breastfeeding behavior won't change your kid's academic outcomes. And yet headlines will continue to tout the correlational findings—potentially misleading people into believing there is a causal link.

Why this obsession with the distinction between correlation and causation in Medicine? Because if A causes B, changing A can change B. But if A is merely *correlated* with B, changing A will have no effect on B. As a doctor, I want things I can act on, things I can change to make your life better. That means I need causal linkages. Correlation isn't enough. Even when correlational data is compelling, it may seem that your doctor is being hesitant and overly cautious about enacting some therapeutic change. That is because we are waiting to have a better understanding of the underlying causal process, if it even exists.

The Biggest Correlation-Causation Disconnect of All Time

Correlational studies often generate the data that supports more rigorous experimental studies, so in many cases we actually know for sure whether observed correlations turn out to be causal or not. For example, it was a correlational study that first suggested that aspirin might reduce the risk of heart attack. In 1950, a family practice doctor, Lawrence Craven, had noticed that aspirin use was correlated with bleeding in his patients who'd had their tonsils removed. Inferring an anticlotting effect, he recommended aspirin to thousands of his patients and reported that they did not suffer heart attacks as frequently as would be expected.

Subsequent randomized trials (aspirin versus a placebo) in high-risk patients would confirm that the correlation truly *was* causal. In patients at high risk, aspirin prevents heart attacks. Or, to put it in the "change" framework, changing certain patients from nonaspirin takers to aspirin takers really does reduce the chance of heart attack. This is causality.

But the path from correlation to causation is not always so easily trod. And I cannot think of a substance with as profound a correlational track record and as weak a causal track record as vitamin D. Vitamin D is a steroid hormone, produced in our skin cells when subjected to sunlight and also contained in certain foods that we eat. It has critical roles in building bone and regulating calcium and phosphate levels. And low levels of vitamin D are correlated with just about every bad thing you can think of. A nonexhaustive list: cancer, dementia, infectious diarrhea, osteoporosis, death from COVID-19, heart attack, stroke, diabetes, and ADHD.

The story with vitamin D has been repeated so often I feel like I know the ending before I get to the results section: Low levels of vitamin D are correlated with a disease state. Researchers, appropriately, test to see whether vitamin D supplementation will reduce the occurrence of the disease state. The supplements are found to cause no change in the disease state.

For example, observational studies found that people with low vitamin D levels were more likely to develop cancer or cardiovascular disease. So a randomized trial (vitamin D supplementation versus a placebo) of nearly twenty-six thousand participants was conducted to prove causality. But there was no difference in cancer rates regardless of whether someone got vitamin D or a placebo. Observational data also showed that individuals with low vitamin D levels were more likely to develop diabetes. A randomized trial (vitamin D supplementation versus a placebo) of more than twenty-four hundred prediabetics found no difference in diabetes rates between the groups. Low vitamin D levels were also associated with falls. So a trial of 2,256 older women was conducted to prove that supplementation would help. It didn't. A Women's Health Initiative study randomized a staggering 36,282 women to vitamin D supplementation versus a placebo to see if vitamin D might prolong life. There was no difference in death rates.

While low vitamin D levels are a sign that you are at risk of various diseases, *changing* your vitamin D level doesn't appear to lower that risk. This is the classic pattern we see when correlation is not causation.

Why do we keep dipping into the correlational well with vitamin D? Several reasons. First, when correlations are strong, we are more likely to assume causation (Tyler Vigen be damned). Second, we have a cheap, widely available, effective, and well-tolerated medicine to increase vitamin D levels; we can take supplemental vitamin D—you don't even need a prescription.

But why are those correlations so strong in the first place? In the case of vitamin D levels, we are (once again, and as usual) in the realm of confounding variables. I have often referred to vitamin D levels as the "lifestyle" biomarker. Let's think of some behaviors that raise your vitamin D level: getting outside, eating a diet rich in fish (and some mushrooms), and exercise (particularly if you do the exercise out in the sun). In other words, your vitamin D level may merely be a marker for a bunch of healthful behaviors you engage in. It also behooves me to mention that darker-skinned individuals will synthesize less vitamin D and are also more likely to experience health inequity.

The relationship between vitamin D levels and all those scary outcomes is *confounded* by those behaviors. When we *randomize* who gets vitamin D supplements in a trial, those behaviors are taken out of the picture. (Since assignment is random, people who get outside in the sun would be equally represented in the vitamin D and placebo groups.)

What these studies suggest is that, sure, vitamin D levels are correlated with bad outcomes, but *changing* vitamin D levels (through supplementation, at least) doesn't change those outcomes. As I've mentioned, the key factor for me is whether a change in A leads to a change in B. But that is because I am a doctor. And to doctors, this is really all that matters.

Finding Exposures You Can Change

If you told me that aging *causes* wrinkling of the skin, I would believe you. After all, there does seem to be a strong association between age and the presence of wrinkles. The two are temporally linked, and there is a clear dose-response relationship. Fine. Aging causes wrinkles. But I don't care. Why? Because I can't change your age. If I can't change the exposure, the causal link between the exposure and the outcome is only of academic interest to me as a physician.

All is not lost, though. Remember that there can be intermediate steps between exposures and outcomes—that's the causal pathway. Yes, aging causes wrinkles. But if we zoom in on that link a little bit, we would find that aging leads to a loss of a protein in the skin called "elastin," which leads to skin sagging that appears as wrinkles. I can't change your age, but perhaps I *can* change your elastin levels. (The quest to find a medication that restores elastin levels is, as you might expect, highly competitive and ridiculously lucrative.)

As a doctor, I am interested in exposures I can change. Smoking causes cancer, and I can help you quit smoking, which means I can reduce your chances of getting cancer. High blood pressure causes heart attacks and strokes, so I try to reduce your blood pressure in order to reduce your chances of getting that heart attack or stroke. I am always looking toward those things we can change together, and I become frustrated whenever causality is assigned to immutable characteristics—like age, sex, or race.

Racial Medicine Is Correlation Without Causation

One area of controversy when it comes to causality is the issue of race. Like vitamin D, race (and, particularly, Black race) shows up time and again in observational data as correlated with a slew of bad health outcomes. But does race *cause* bad health outcomes? You could argue that, by my paradigm, it is a moot point. I can't change your race, so the presence or absence of a causal link is immaterial. But I feel like that is letting me off the hook too easily here.

First, it's important to realize that biologically speaking, race is a nearly useless metric. Most people assume that people of different racial groups have substantially different DNA, but this is not actually the case. For virtually any gene you look at, the variation *within* a racial group is as wide as the variation *between* racial groups. That means using a racial group to predict whether a patient has a gene that may, for example, confer resistance to a certain blood-pressure medication is a terrible idea. Rather, you'd get a lot more mileage from just measuring the gene itself and not bothering with race at all. From a genetic perspective, humans are incredibly homogeneous; any two random humans differ from each other by about one in one thousand DNA letters, substantially less than the variation within our closest relative, the chimpanzee. The belief that race is an important biologic factor occurs because race is something we feel capable of judging from a distance. But, as the saying goes, it is truly just skin-deep.

So if race can't be changed, and race is not even an important biological construct, why do Black people seem to have worse outcomes across a variety of key health metrics, ranging from maternal mortality to overall life expectancy? The answer is racism. Racism lies along the causal pathway from race to bad health outcomes. And, no, I don't only mean the overt racism that leads to disturbing statistics like ten in every ten thousand Black men will be killed by police compared to four in every ten thousand white men. I mean the ongoing force of societal marginalization that leads to poor access to preventive healthcare, poor access to nutrition, more limited educational opportunities, and exposure to more violence.

Correlational studies, where race is used as a predictor for a disease state, have a tendency to distract from the underlying problem. We may see a study that notes that Black people are twice as likely to develop diabetes as white people and, erroneously, concludes that it is due to some inherent *unchangeable* biology, when in fact this is a correlation induced by multiple third factors—confounders such as poor socioeconomic conditions, which are things we *can* change.

This is why I've moved my lab away from using race as a variable in our statistical models. It's not that there is no correlation between race and the kind of stuff I research (kidney disease outcomes). There is. But race is correlational, not causal. Better instead to focus on the real causal agents: racism (implicit and explicit) and societal inequality. While I don't have a pill to fix those, I am fortunate enough to have a platform in which to urge everyone to recognize the causality of health inequality in this country and to move our government and ourselves toward addressing it.

The Gold Standard for Causality in Medicine

While Sir Austin Bradford Hill gave us nine criteria for causality, when it comes to medical research, experimental evidence reigns supreme. In part, this is because so many of Hill's other criteria are subjective or difficult to measure, whereas the results of an experiment can be tightly controlled. The gold standard of experimental evidence in Medicine is the "randomized controlled trial" (a paradigm we will explore more deeply in chapter 5).

Randomized trials take a group of individuals and assign them, at random, to a treatment of interest and either a comparator treatment or a control (which is often a placebo medication, or sometimes usual care). Randomized trials interrogate causality because they explicitly test the central idea of causality: If A causes B, then people randomly assigned to A should have more B. If consuming olive oil truly causes a longer life span, then people randomized to consume olive oil should live longer than those randomized to not consume olive oil.

The main benefit of randomization is that it deals with confounders. Remember that for a variable to confound an exposure-outcome relationship, it has to be associated with *both* the exposure and the outcome. In a randomized trial, *nothing* is associated with the exposure because the exposure (drug versus a placebo, for instance) is chosen at random.

Randomization balances characteristics between the treatment and placebo groups. By randomizing people, we can be sure that a similar proportion of men and women will end up in each group, for example. And this holds true for things like gender, but also for every other potential variable—*even things you haven't measured.*

In the nonrandomized, observational framework, we can "adjust" for certain confounders. If we find that a certain drug is taken by older people preferentially, we can use statistics to adjust our analysis for age, removing age as a potential confounder of the relationship. What we can't use statistics to fix are *unmeasured* confounders. If you don't know whether the individuals in a study smoke or not, there is no way to adjust for smoking. And however creative you are, you'll never think of every possible confounder. When I teach about confounding in the med school, I often talk about mojo. Sure, you can adjust for age, sex, smoking status, blood pressure, history of hair loss, whatever, but certain people just have more mojo than others—and you can't measure that. There's always another confounder out there.

But not if you randomize. If you pick the exposure at random, you ensure that even people with a lot of mojo are just as likely to get the drug or the placebo. Randomization doesn't just balance all your observed confounders; it balances your *unobserved* confounders as well. That eliminates confounding, meaning that *if* a correlation is observed, you can be that much more certain it is a causal one.

Marijuana and the Street Price of Cocaine

Of course, not every hypothesis lends itself to a randomized trial. Let's say I want to know whether marijuana is a gateway to harder drugs, like cocaine. In other words, does the use of marijuana *cause* the use of cocaine in the future? This is an important question because if there is a causal link here, changing people's marijuana-smoking habits would be expected to change their subsequent cocaine-usage habits. (This would have major implications for the marijuana legalization movement.)

The correlational data is pretty compelling. According to the 2013 National Survey on Drug Use and Health, which surveyed around fifty-five thousand teenagers, those who reported smoking pot were about eighty times more likely to have reported trying cocaine than those who did not smoke pot. In fact, of the thousands of people in the data set, there were only 135 who reported trying cocaine but *never* smoking weed. Still, that is just a correlation. There are plenty of potential confounders here. Maybe people who are inherently less risk-averse are more likely to use any given drug. Maybe people who feel socially isolated are more likely to use both marijuana and cocaine. I'm sure you can think of other confounders.

The easiest way to tease out causality here would be to do a randomized trial. We'd simply enroll one thousand people, randomize half of them to smoke marijuana frequently and the other half to do nothing (or maybe to smoke a placebo joint) and wait a few years to see who ends up using cocaine.

There is obviously a problem with this study design. It is unethical—offering no potential benefit to those who enroll, and substantial risk. We can't do it. This is the case for many important medical questions. We can't randomize people to smoke cigarettes, get pregnant, hold their breath for three minutes, or even consume large quantities of bacon. For much of the past fifty years, in those situations we've had to rely on observational data (with all its flaws) and squint at correlation until we decide it is causation, accepting the risk that we may be completely incorrect.

But some novel techniques are offering new hope for causal inference. The key is to find a variable that is associated with the exposure of interest, but *not* the outcome. These are known as "instrumental variables," and though they don't provide all the benefits of randomization, they may be the next best thing.

Just as randomization is linked to the exposure (perfectly linked, since you get whatever you were randomized to) but not the outcome (except via the exposure), so are instrumental variables. A key instrumental variable to answer the pot question is the price of pot. In 2009, researchers published a paper trying to answer the question of whether or not pot was a gateway to harder drugs like heroin. Their insight was brilliantly simple. The price of marijuana is directly linked to marijuana consumption—the higher it goes, the fewer people smoke. But the price of marijuana should not affect how often people buy cocaine. (Cocaine and marijuana prices are not related.)

As such, if there is a correlation between the *price* of marijuana and the *use* of cocaine, it can be inferred that the causal pathway must flow *through* marijuana usage. If marijuana is truly a gateway drug, when marijuana prices drop, cocaine use should increase and vice versa. If marijuana is not a gateway drug, then the price of marijuana and the use of cocaine should be uncorrelated.

The results were pretty clear: Overall, there was no sign of a gateway effect, no link between marijuana prices and cocaine usage. To be fair, the researchers did identify a group of "troubled youths" for whom the opposite appeared to be true—it may be that marijuana *is* a gateway for some people, but not for most of us. Fair enough.

Building Trust in the Absence of Causation

Let's face it—there is a power imbalance in the doctor-patient relationship. You are in charge of your own body. There is no way I can force you to take a medication that you don't want to take, nor would I want to. But the converse is not true. There are a slew of medications and treatments a patient may want to take, but they are not given that opportunity unless they get a prescription from a doctor. We are not the sole arbiters of scientific truth, but we are the ones with the prescription pad.

One can imagine a truly libertarian world where medications could be dispensed without prescriptions, and the role of the doctor would be to merely provide advice about which were good choices, but that is not the world we live in. I would argue we actually wouldn't want to live in that world, given the complexity of medication side effects, interactions, and effects on individuals that doctors learn throughout their training, but given that the only person you would be harming by omitting or ignoring a doctor's advice is yourself, I admit I get the victimless-crime aspect of the argument.

So how do we maintain trust when the desires of a patient for a treatment and the willingness of a physician to prescribe that treatment are in conflict? There needs to be movement on both sides. Patients need to understand *why* we are so hesitant to prescribe medications, even benign ones, that don't have a strong causal evidence base. It would essentially leave us completely adrift—with no signposts to know whether we are helping or hurting, and no rationale to choose one particular therapy over any other. We simply can't practice Medicine that way. Patients need to resist the urge to "doctor shop" to find a doctor who will do what they want him or her to do. If you find yourself in a situation where only one out of six doctors will agree to the course of therapy you are interested in, you need to consider why that might be the case.

On the other hand, doctors need to understand that causality is *not* black-and-white. In fact, we can never be 100 percent sure that two things are causally linked. Even the lightbulb turning on when we flip the switch could be a wildly improbable set of coincidences, after all. We need to say no, clearly and consistently, when the risks of a treatment outweigh potential benefits. And we need to explain *why* we are saying no, referencing our oaths if necessary. We need to suggest alternative treatments with a stronger evidence base, answer questions, and be willing to explain how correlation and causation are different. We need to listen to our patients, to their experiences with past treatments and their concerns about future treatments, since all treatment is, in the end, individual. In some circumstances, to maintain the bond of trust, we can even consider a time-limited trial of a therapy (provided there are no major risks). The key here is to set the guidelines up front: How long will we do this thing? And what do we agree is the yardstick by which we will measure success or failure?

In the end, the goal of Medicine is to change something—to make you healthier and happier, to cure your illness or reduce its burden. To change an outcome, you need a causally linked exposure. If A causes B, changing A changes B. That is the central paradigm of medical research and practice. And it is the reason that the quest for causality continues.

CHAPTER 5

How Coin Flips Changed Medicine Forever

THE INITIAL PEAK of the COVID pandemic in Connecticut occurred in April 2020. My hospital, Yale New Haven, had one thousand beds, and five hundred of them were filled by individuals with COVID. The intensive care units full, we had converted several floors to serve as temporary ICUs and were pulling nurses and doctors from all over the health system to run them. I was drafted to be the attending physician on one of the COVID wards for patients who were not yet critically ill. Being the physician of record—the one who is ultimately responsible for all the medical decisions—is a job I'm quite comfortable with, but this experience was fundamentally different from any I'd had before.

For one, the usual white-coated crowd that makes up the academic medicine team—attending, fellows, residents, students—was absent. I was teamed with only one junior physician, a fellow from the kidney disease program. The medical students were learning remotely, the residents pulled to other assignments. The ward, though full of patients, felt lonely, almost empty. No visitors were allowed, so the usual sounds of families coming, going, and chatting with their loved ones were absent, replaced only by the intermittent beeping of IV poles and the punctate coughing of infected patients.

The formidable silence would be broken once or twice a day by an alarm, followed by a "rapid response" call. A step before the dreaded "code blue," which would imply that CPR is urgently needed, the rapid response is designed to get a patient stabilized before the worst can happen. On the COVID ward, this meant a patient who had respiratory decompensation, their oxygen level falling below what our area of the hospital was equipped to support. They would be rushed to one of our improvised ICUs for more aggressive intervention, typically involving intubation and mechanical ventilation. Luckily, as statistics would later show, only around 20 percent of the patients admitted to the hospital would require that level of care. The majority of those on my ward needed some support—nasal oxygen, Tylenol, anti-nausea medication—but after about a week, they would be able to return home, worse for wear but alive.

I discovered quickly that my ability to predict who would be among the 80 percent that make it home after a week and who would be part of the 20 percent that go to intensive care was abysmal. One of my first patients on the ward was a ninety-three-year-old man, admitted from a nursing home. The data would suggest he'd have a terrible outcome. He didn't. Within a few days of his arrival on the ward, I entered his room to find him calmly reading the newspaper, no longer requiring oxygen. He left the hospital two days later.

And then there were the young and healthy people, like a thirty-five-year-old man who, despite having no comorbidities, relentlessly worsened day after day, almost like clockwork. On day 1, he had a two-liter oxygen requirement. Day 2, four liters. Day 3, six liters. By day 4, he was in the ICU. (He would survive, though he had a month-long hospitalization requiring ventilator support for much of it.)

Walking into those rooms always gave me a dizzying sensation, like I was approaching the edge of a cliff. So it was with Ms. Jepson, a licensed practical nurse who had worked in several nursing homes in the area. Nursing homes had been hard hit—with death rates approaching 25 percent—and, despite her best efforts and using what personal protective equipment she could get, she became infected. Her infection led to infections in her husband and their son (age eighteen), and while the son was able to recover at home, her husband was in the ICU while she was on my ward.

In April 2020, we really had no idea how to treat COVID. This was before vaccines were available, of course, but also before remdesivir, or the trials that confirmed that certain steroids worked, and before we had data on monoclonal antibodies. We weren't even entirely sure how easily COVID spread in the air versus on surfaces. We were essentially flying blind. And we made some bad choices because of it. Due to the observation that some of these patients declined extremely quickly, we had adopted a policy of "early intubation"—putting people on the vent once the oxygen requirements crept to a concerning, but not exactly life-threatening, level. Future studies would teach us that a more permissive strategy—trying to hold off on intubation for longer—led to better outcomes. But at the time, when the oxygen requirement got to about six liters, we were told to call the ICU to get the ball rolling on the ventilator and prepare the patient for what was coming next.

Ms. Jepson flirted with that six-liter number multiple times over her first week in the hospital, hovering just under it for days. Knowing how rapidly decompensation can occur, I arrived to the hospital every morning convinced that the night team would confirm my fears and tell me she had left our ward for the ICU. But day after day, she hung on, as brave as any patient I've known.

Though we didn't know how to treat COVID, it didn't stop us from trying some things. Ms. Jepson received hydroxychloroquine, and also tocilizumab—a specialized antibody that could reduce inflammation. Hydroxychloroquine was part of Yale's official COVID-19 treatment algorithm at the time, based on some observational studies that suggested the drug could be effective. From my front-lines perspective, I wasn't impressed. It seemed regardless of what I did, 80 percent of my patients would get better, and 20 percent would go to the ICU, and hydroxychloroquine didn't move anyone from one column to the other.

Ms. Jepson had a rough time with the drug, in particular. Given its potential to lead to abnormal heart rhythms, we monitored her EKG daily—prompting us to hold the med on several days because of signs that the electrical impulses driving her heart were being adversely affected. But for her, the real problem with hydroxychloroquine was the nausea, which made her less able to eat and drink, and sapped her of strength. The days we held hydroxychloroquine due to the abnormal electrical signals in the heart were, ironically, her best days.

She asked me more than once if the drug was really necessary. I told her the truth, which was that I didn't know. I told her that it was plausible that the drug could work, but we didn't have adequate data yet. Given the severity of the pandemic, though, I said, it was worth a try. Ever the good patient, Ms. Jepson took her medicine on her roiling stomach and continued to hope for home.

I'm happy to say she was one of the 80 percent. She made it home after just over two weeks in the hospital, her husband a few weeks after that, to their son, who had been left alone for the first time in his life while both his parents fought for theirs.

The provision of hydroxychloroquine without solid evidence turned out to be a mistake. Larger studies, including randomized trials of the drug, eventually confirmed that the medication had no effect in reducing severity of or mortality from COVID-19. In fact, most studies reported increased adverse events, including serious adverse events, among those randomized to receive hydroxychloroquine compared to those who received a placebo. These included heart rhythm problems, and kidney and liver failure.

While we may not ever have a full accounting of the number of people who were harmed by the wide-scale use of hydroxychloroquine early in the pandemic, we know now that there is no benefit, and we can consider any side effect to be an unwarranted harm. At the time, though, I was happy to have something—*anything*—to prescribe for the novel virus. And, as history would bear out, some of those early guesses (like tocilizumab) turned out to be decent. But in the moment, what else could we have done? Should we have sat back and done nothing?

What we could have done is a randomized controlled trial. We could have been open and honest with our patients and the public, and said to them, "Listen, this drug may work, it may not, and it may harm you. The only way we'll learn if it will work is if we randomly give it to some people and randomly don't give it to others."

If, instead of prescribing hydroxychloroquine as a "best guess" medication, we had given it at random, we could have completed a large-scale definitive trial of the drug across the United States in under a month. By May 2020, we would have had the answer. If it had proved to be a miracle drug, we'd have known for sure and could have provided it to everyone else. If it had been a dud, we'd have known for sure and could have avoided the cardiac toxicity and nausea in all of those we would have otherwise treated.

Unfortunately, we didn't do this, and, to be fair, we *couldn't* do this, because the structures to actually conduct a clinical trial are so complicated. It frustrates and baffles me that it is far, far easier in Medicine to simply *do* something—treat a bunch of patients with a medication—than to *study* something. And the space between those two actions can be measured in lives.

THE HISTORY OF Medicine has been marked by some turning points so dramatic it is difficult to imagine how Medicine was practiced before they occurred. Before "germ theory" was developed, for example, there was no consistent, verifiable explanation for how infectious diseases occurred or why certain treatments (like dousing a fresh wound with alcohol) seemed to be effective. After Louis Pasteur fully fleshed out the relationship between microscopic organisms and disease in humans, the world was forever changed. Infectious disease—far and away the number one killer of human beings for the vast majority of human history—is now responsible for just a small fraction of global deaths. Tuberculosis, the number one infectious killer of humans in the world prior to COVID-19, led to 1.4 million worldwide deaths in 2019. About eighteen million people died of heart disease in that year.

The development of surgical anesthesia, antibiotics, and vaccination no doubt belong with germ theory among the earth-shaking turning points that led to modern medicine. But one turning point trumps them all: the randomized controlled trial. Without RCTs, modern medicine would not exist. They are the knowledge-generation engine of Medicine, and as Medicine becomes more complicated, they become more critical by the day.

The idea of a randomized trial represents a fundamental shift from traditional medical practices, which focus on the relationship between an individual doctor and an individual patient. Under ordinary circumstances, a doctor formulates a treatment plan based on the characteristics of a specific patient: the patient's risk factors, risk tolerance, disease status, other medications, and a host of other factors born out of long experience. The patient and doctor discuss the recommendation, modify it, and, eventually, implement it. Treatment is personal.

Randomized trials, in contrast, are fundamentally impersonal. Indeed, that is perhaps their greatest strength. By choosing who is treated *at random*, randomized trials remove all the biases and assumptions that go into typical treatment decisions. What's more, by choosing who is treated at random (and carefully noting the outcomes of treatment), randomized trials ensure that the group of treated individuals and the group of untreated individuals are similar in most ways. There is no other place in Medicine where this is the case. It makes the randomized trial the most appropriate and ultimate arbiter of scientific proof, and also the least personal enterprise of modern medicine.

To understand how Medicine works, we need to understand how randomized trials work. For at this point in history, they are inextricable. This involves understanding what makes randomized trials so special in their design, execution, and analysis. It involves understanding what to make of a trial that shows a positive result, and what to make of a trial that shows a drug may be ineffective. It involves knowing how even the vaunted randomized trial can get it wrong—for as with much of Medicine, our tools are very good, but far from perfect.

Proof

It was 1747, and if you were a sailor in the British fleet, the most powerful in the world, there was a good chance that you would die of scurvy. Just a few years earlier, Commodore George Anson had set off to circumnavigate the world with nineteen hundred men. Fourteen hundred died, the vast majority due to that disease. But at the time, the array of symptoms was baffling.

Physicians then did not know that scurvy was caused by a deficiency of vitamin C. In fact, the concept of "vitamins" (small molecules essential for life) had not been invented yet. Vitamin C is necessary for the proper synthesis of collagen, the main protein that forms the connective tissue—the scaffolding of the body. With inadequate collagen to support them, small blood vessels ruptured, leading to bleeding gums and bruising. Collagen forms part of bone; deficiencies led to fractures and loose teeth. Wounds didn't heal properly. Deaths were common, with one estimate suggesting that two million sailors died of scurvy from the fifteenth to the eighteenth century.

Necessity being the mother of invention, multiple cultures around the world stumbled upon various cures for scurvy. The Chinese carried ginger on their ships to combat the disease. The Native Americans boiled needles of the white cedar tree to create an antiscurvy tonic. In the late sixteenth century, the English explorer Sir Richard Hawkins recommended drinking orange and lemon juice as a scurvy cure. The common thread among all these treatments, we know now, is vitamin C. But at the time, they were more or less discovered by trial and error. Since vitamin C reverses scurvy symptoms so rapidly, attributing causality was fairly straightforward. Nevertheless, various treatments for scurvy, both effective and ineffective, remained in common practice for centuries, without a clear understanding of which were truly beneficial.

While certain physicians advocated for sailors to be given citrus fruits, world navies remained unconvinced. Lemons and limes were expensive, and provisioning the entire fleet with adequate stores of citrus would always take second place to that more critical supply: grog, an alcoholic staple of the British navy that would not spoil on long sea voyages.

All of that would change in the middle of the eighteenth century, when James Lind, a Scottish doctor, did something no other doctor had thought to do. Instead of relying on anecdote and experience, he decided to generate some hard evidence. In what is possibly the earliest reported design of a clinical trial, Lind identified twelve sailors with scurvy and divided them, presumably at random, into six groups of two. (As an aside, this is quite an ambitious trial design.) Two received citrus fruit, two dilute sulfuric acid, two a quart of cider, two vinegar, two seawater, and two barley water. Lind believed (incorrectly) that scurvy was due to decay, or putrefaction, of the body. He thought acids (like those in vinegar or citrus fruit) could stave off that decay. He reported dramatic results. The two sailors given the citrus fruit quickly recovered; the others continued to languish with the disease.

Over the next several decades, evidence generated by Lind and other physicians helped cement the role of citrus fruits in the treatment and prevention of scurvy. The British navy eventually came around and by 1800 was issuing lemon juice to all of its sailors. At the time, the terms "lemon" and "lime" were used interchangeably, and the American derogatory term "lime-juicer," later shortened to "limey," was born. The last laugh would be with the British navy, however, who, thanks to their reliance on evidence, had some of the healthiest sailors in the world for more than a century.

Lind's study represented a true paradigm shift in Medicine. It was the first time a physician-scientist had decided to test the courage of his own convictions. For years—centuries, even—the idea that citrus fruits, or fruits in general, or "healthy eating" might be the treatment for scurvy had percolated through the medical intelligentsia, but Lind actually put it to the test in a controlled manner.

I'd like to tell you that he was rewarded for this revolutionary act, that he was lauded by his contemporaries as the visionary he was. But like so many visionaries, he went relatively uncelebrated in his seventy-eight years on Earth. Still, his act was the start of something—the most rapid advancement of human health in history. Lind had designed the randomized controlled trial—the gold standard for medical evidence from that point onward. The RCT is what separates modern medicine from ancient medicine. There are innumerable ways to design a medical study, but RCTs are so unique, so powerful, that some doctors will refuse to change a medical practice without their support. I am one of them.

My Obsession

For the past five years, I've dedicated a chunk of time every week to discussing a new medical study in a public forum. While directed at healthcare providers, these Impact Factor reports have become somewhat popular online, which is gratifying but comes with quite a bit of baggage in the form of (mostly) anonymous trolling and harassment. I have been called a liar, a shill, not a real doctor, a Russian plant, a Chinese plant, a sociopath, a murderer, a fool, and "the reason we are going to lose our democracy."

My pieces are not particularly inflammatory. I do my best to report on the data impartially and point out how that data might be misinterpreted or misapplied. Of course, my personal opinions and biases shine through. (You are already well aware of my inherent skepticism toward "one simple thing" studies and miracle cures.) But mostly what I do in those pieces is point out that there might be a kernel of interesting data here but that (usually) we need stronger evidence to make firm conclusions. My writings that upset people the most are those that contravene their motivated reasoning—those that question a conclusion that someone wants, or needs, to be true.

Questioning conclusions into which individuals have invested their time, social capital, and identity is a recipe to create anger. Earlier in the book, we learned that people will go to great lengths, even inventing new data, to defend conclusions they want to be true. This is why some folks accuse me of secretly being a Russian or Chinese agent; it makes the discounting of my opinion logical. To me, it is not controversial to say that a randomized trial is needed to evaluate a particular intervention. And yet, particularly if the intervention is supported by a passionate group of people, even the act of asking for high-quality evidence in the form of an RCT can fill my email box with insults and threats.

Every study is different, but, almost invariably if I am talking about a study that is *not* randomized, I will mention that in the absence of randomization, those pesky confounders (measured and unmeasured) limit our ability to assess causality. For example, in 2019, *JAMA Pediatrics* published a study linking higher maternal urinary fluoride concentration to lower IQ in children. Since urinary fluoride is a proxy for fluoride intake, if the observed relationship had been *causal*, this study would have had profound implications. It would have suggested that we were systematically lowering children's intelligence in service of preventing some cavities through the process of water fluoridation. In other words, this would have been a travesty.

But this was not a randomized trial. Pregnant women were not assigned at random to take fluoride tablets or drink from a fluoridated versus nonfluoridated source. This was an observational study, so it was subject to measured and unmeasured confounding of the sort we discussed in chapter 4. While ingesting more fluoride does increase the amount of fluoride in your urine, it is not the *only* thing that increases fluoride in your urine. More concentrated urine will have more fluoride in it, for example. Less acidic urine, in fact, will have more fluoride in it (as acid promotes the reabsorption of fluoride in the kidney).

As such, the study *may* have found that fluoride reduces IQ. Or it may have found that women who drink more water have smarter babies. Or that those who eat more meat (which leads to acidic urine) have smarter babies. Or it could be something else that none of us have thought of yet. A nonrandomized study can rarely provide the certainty we need to make informed medical decisions.

The typical rejoinder to my "This isn't a randomized trial" line of thinking comes in two flavors. In one version, critics tell me that the study was good enough—the data is strong enough that a randomized trial is not necessary. Another version goes deeper, asking me, to quote one YouTube commenter, "Why are you so *obsessed* with randomized trials?"

I will admit it: I am obsessed with randomized trials. In my mind, there is no better way to determine whether therapies will help people or not—to determine causality. I understand the visceral response to them, though. I understand that the process, randomly assigning a string of patients to a drug or a placebo, feels crude and, worse, cold (despite the fact that patients give consent to become trial participants). Isn't that the antithesis of the doctor-patient relationship? Instead of having a discussion of risks and benefits and goals of care, I am going to flip a coin and do whatever that coin says? How can that possibly lead to good medicine?

The fact is, randomized trials aren't Medicine. They are scientific tools that *lead* to Medicine. Just as an x-ray finds a broken bone, a randomized trial finds broken treatment paradigms or broken standards of care. We don't fault the x-ray for being cold, unfeeling, and singularly obsessed with the detection of cracks in bones. Nor should we fault the clinical trial for its laser-like focus on a single, discrete medical question.

Do these trials ignore the broader picture of the patient? Their wants, needs, and desires? Absolutely. Do we have a better way to determine which medicines work? Not even close.

The Secret Sauce

There is more than one way to conduct a randomized trial, but all randomized trials have the same special ingredient—the same secret sauce, as I tell my students—the coin flip. The act of determining the treatment assignment at random, as opposed to by what a fallible, biased human might want to do. Nowadays, we don't really flip coins to decide who gets an experimental treatment and who gets a placebo; computers do the coin flipping for us. But the idea is exactly the same. By assigning the treatment at random, we disentangle the choice of treatment from the patient characteristics, allowing us to evaluate whether the treatment works without being confused by the reasons we might give the treatment to a patient.

As an example, would it surprise you to hear that the risk of heart attack among people taking cholesterol-lowering medications is, dramatically, several-fold *higher* than the risk of heart attack among people who don't take those medications? Taken at face value, you could look at this data and conclude that these drugs should immediately be banned.

But in actuality, this data isn't worrying at all. Here is a list of people who don't take cholesterol medication: children, those without heart disease, those without high cholesterol. Here is a list of people who do take cholesterol medication: older people, people with heart disease, people with high cholesterol. In other words, doctors *select* who to treat based on their risk factors and, appropriately, give these drugs to people at higher risk.

Looking at the observational treatment data, it is impossible to tell whether the drugs are causing heart attacks or whether it's simply that the drugs are prescribed to people who are already at risk of heart attacks. But by randomizing, we balance all those covariates. Randomization ensures that the treatment and placebo groups have similar ages, similar cholesterol levels, and so on. It is the ultimate causality-confirming machine.

That is why you see randomized trials published in the most famous medical journals—the ones your doctor is likely to be reading—while observational studies are frequently relegated to the lower tier. It's also why randomized trials generate headlines.

But just because randomized trials offer the best evidence we have to prove that a medication works doesn't mean they are perfect. Like any study, randomized trials can be misinterpreted. And often, that misinterpretation turns on a single word: "significant."

Not Significant the Way You Think

If an event is significant, it means it is important. If a person is significant, we take note of them. Significant places evoke our emotions and nostalgia, our fondest memories. No one, unless they take statistics classes, thinks "significant" means "data that is a bit more extreme than what would be expected, assuming there is no underlying process driving the data to that extreme." Yet that is exactly what "significant" means in every medical paper.

And when those papers are covered by the press, that word is frequently lifted verbatim: "Researchers confirm a significant effect of minoxidil on hair growth." Presumably, that drives more engagement than "Researchers obtain data that would be rather unusual if minoxidil does not have any effect on hair growth." That one word—"significant"—may have done more damage to how doctors and patients understand medical research than any other.

So what *do* researchers mean when they use that term? The results of a study are either significant or not significant. Convention dictates that this is a binary classification scheme, implying some threshold beyond which a researcher has the right to claim significant findings. That threshold is known as the "p-value," one of the most misunderstood but important concepts in medical statistics.

Medical research papers test a hypothesis, but because of some vagaries of math, they test what we call the "null hypothesis"—the hypothesis that whatever you are trying *won't* work. Imagine that I have a new drug to treat Ebola, a deadly hemorrhagic fever virus. To prove that the drug works, I randomize a group of Ebola sufferers to the drug versus a placebo. I cross my fingers and wait.

How many in each group will survive? If more die in the placebo group, can I conclude the drug works? That's a bad idea. Even if the drug you are giving does nothing at all, it would be incredibly unlikely for *exactly* the same number of individuals to die in the treatment and placebo groups. That's just the randomness of the universe at play.

Consider a placebo versus placebo trial: If I enrolled one thousand individuals with Ebola and gave half of them a placebo pill marked "A" and half a placebo pill marked "B," would we expect the exact same number of people to die in both groups? Of course not. We'd expect a *similar* number—I gave both groups a placebo, after all—but because of random chance, one group would have more deaths and one group would have fewer deaths. Similar but different outcome rates should not, therefore, impress us very much.

But what does "similar" mean? We can grasp it intuitively. We know, for example, that in a study of one thousand people, with five hundred in each treatment group, fifty deaths in the treatment group compared to fifty-one in the placebo group is not very impressive—those are similar numbers. But ten deaths in treatment versus fifty-one in placebo? That seems interesting. Results like that would be pretty odd if the drug had no effect at all.

The measure of that oddness, of how weird the data is assuming the drug *doesn't* work, is quantified in the p-value. Specifically, the p-value is a measure of how unusual the data you observe is, under the assumption that the treatment of interest has no true underlying effect. The lower the p-value, the weirder the data is. If it's low enough (and the commonly but arbitrarily agreed-upon threshold for this is 0.05), we reject the hypothesis that the drug has no effect, de facto embracing the idea that the drug *does* have an effect. We call this "statistical significance." A p-value of 0.05 means that the data you see (or even weirder data) would arise only 5 percent of the time if the drug didn't work at all.

There is nothing magical about p=0.05. We could have chosen any cut point for statistical significance. We *could* have chosen no cut point at all. To give some visceral intuition of what p=0.05 means, think of an ordinary quarter. Imagine you flip the quarter four times and get four heads in a row. That's a bit weird, but obviously not impossible. Doing some math would tell you that such an event would happen only 6.25 percent of the time. Five heads in a row happens only 3.125 percent of the time. So a p-value of 0.05 is about as weird as flipping a coin and getting somewhere between four and five heads in a row. In other words, not *too* weird.

If it happened to you, you probably wouldn't conclude that you had some trick quarter on your hands—you'd just think you got pretty lucky. And yet that is the threshold that the medical establishment, including the FDA, uses to evaluate a new drug. Is the observed effect of the drug fairly weird assuming the drug doesn't work? Yes? Well, then the drug probably works.

That arbitrary threshold of 0.05 has created something of an obsession in scientists, and it's easy to see why. If I do a trial of my Ebola treatment and arrive at a p-value of 0.06, what I am saying is that data as weird as this would happen only 6 percent of the time, assuming my drug doesn't work. If I arrive at a p-value of 0.04, I am saying that data as weird as this would happen only 4 percent of the time, assuming my drug doesn't work. And yet I can describe only the latter as significant. The latter would lead to headlines that read NEW DRUG SIGNIFICANTLY REDUCES THE RISK OF DEATH FROM EBOLA, while the former (the 6 percent example) would lead to headlines that read NEW DRUG HAS NO SIGNIFICANT EFFECT ON THE RISK OF DEATH FROM EBOLA.

That tiny difference—two percentage points—changes entirely how doctors, patients, and the public at large understand a scientific study. Speaking from experience, I can tell you that there are few things more depressing for a scientist than plugging your data into statistical software and seeing p=0.06 come out.

Things get a *lot* more concrete once you have two studies with a p-value less than 0.05. Or three, or four, or more. As the data mounts, you can become much more certain that your drug really does something. It's like if you took that quarter with five heads in a row and gave it to your friend, and *he* got five heads in a row, and he gave it to his sister and *she* got five heads in a row. At this point, you can be fairly sure that this isn't your typical twenty-five-cent piece.

What this means is that any single clinical trial is rarely (perhaps never) definitive. Granted, some studies have results that are way weirder than p=0.05. I have seen studies where the data is so weird that it would happen only one in a thousand times, or one in ten thousand, assuming the drug is a dud. With 2.5 million research papers published a year, of course some very weird outcomes will occur. Our greatest weapon against random weirdness in data is replication.

So what makes a drug *actually* significant? The key thing is to distinguish between "statistical significance" (with its p-values and null hypotheses) and "clinical significance." If something has clinical significance, it means its effect is relevant to a real human. Statistical significance is objective; clinical significance is subjective.

For example, imagine a new blood-pressure drug hits the market. A rigorous randomized trial confirms that the new drug lowers systolic blood pressure by one point compared to a placebo. We enrolled tens of thousands of people in the study, and that difference was highly statistically significant at p=0.01. That means that the data is fairly weird—we'd see a difference of one point of blood pressure just 1 percent of the time in a study this large, assuming the drug had no effect. We conclude that the drug does indeed have an effect on blood pressure. But would I use the drug? Absolutely not. One point of blood pressure is irrelevant when it comes to real humans. I don't need to get people from a systolic pressure of 160 to 159. I need to get them below 140. This drug is not clinically significant.

What this boils down to is that just because a randomized trial is "positive"—it has statistical significance—does not mean the drug is good, worthwhile, or recommended. In order to get to that point, you need multiple trials, as well as evidence of benefit in terms of hard outcomes like death or quality of life.

But what about the converse, the so-called "negative" trial, which does not reject the null hypothesis? The trial that says "Yeah, the data we see here is entirely consistent with the idea that the drug doesn't work." Even in this case, we need to be cautious with interpretation.

"Negative" Isn't Always Bad

The secret sauce of a randomized trial is that coin flip, ensuring that all the little factors that make us *us*—our age, our blood pressure, our mojo—are relatively balanced across the two treatment groups. But the coin flip is not enough. The second ingredient of a good randomized trial is participants. Specifically, the number of participants.

Let's say I am conducting a trial of an often-overlooked condition: boredom. I randomize ten participants with severe boredom to either go play dodgeball or go play tetherball. Of the five who played dodgeball, two recover from their boredom. But three recover from their boredom in the tetherball group. Do you feel confident that tetherball is the superior treatment here?

I don't. There just weren't enough people in my study to rule out the vicissitudes of chance. And if I calculated a p-value, it would confirm this. The math works out to a p-value of 0.53, suggesting that results this weird would happen more than half the time, assuming dodgeball and tetherball have no different effect on boredom.

Now, if I did the same trial with ten thousand bored folks and cured two thousand in the dodgeball group but three thousand in the tetherball group, I would feel much more confident recommending tetherball over dodgeball to my bored patients in the future. (The p-value in that trial would be less than 0.0001.)

Note that both of these hypothetical studies had the same proportion of response in each group but vastly different interpretation. Why? Because sample size—the number of participants in a randomized trial—matters.

We can think of the sample size of a randomized trial as analogous to the power of a microscope. A truly gigantic trial may be able to pick out tiny clinically insignificant differences in treatment effects (such as my hypothetical blood-pressure drug that truly reduces blood pressure, but only by one point). Conversely, a small randomized trial is a weak microscope, able to detect only truly huge treatment effects (like the effect of vitamin C in scurvy).

When the sample size we use is too small to detect clinically meaningful treatment effects, we consider a study "underpowered." This means the study doesn't have the ability—the power—to detect an important effect, because there aren't enough people to wash away all the chance findings that make the study of human health and disease so difficult. Underpowered trials happen all the time and are frequently misrepresented in the media as definitively negative—proving no effect—when in fact they are merely proving no *large* effect.

For example, in 2014, the results of a randomized trial that tested a compelling hypothesis were published in the journal *Diabetology & Metabolic Syndrome*: Could melatonin cure metabolic syndrome? "Metabolic syndrome" refers to a group of findings that tend to cluster together among individuals who are overweight: elevated blood sugar, high cholesterol and triglycerides, enlarged waist circumference, and high blood pressure. Hundreds of studies have shown that those with metabolic syndrome are at dramatically elevated risk of developing diabetes or having heart attacks or strokes. A cure would be transformative. Melatonin is a hormone produced in the pineal gland, and studies in mice with metabolic syndrome had shown that melatonin supplements could improve several of the metabolic syndrome parameters. Would the cheap, widely available supplement work in humans?

To find out, the researchers randomized thirty-nine individuals with metabolic syndrome to receive melatonin or a placebo. After melatonin supplementation, 35 percent of participants were free of metabolic syndrome, while just 15 percent of those taking a placebo had that outcome. This is a fairly impressive result on the surface—implying that one would need to treat just five people with melatonin to cure one person of metabolic syndrome. This is a trade most doctors would make in a heartbeat.

But the finding was not statistically significant. The p-value was 0.25, meaning that results as strange as these (a 20 percent difference in outcome rates) could be expected to happen one-quarter of the time in a trial of this size, assuming melatonin has no effect on metabolic syndrome. Thus, the study was reported as negative, and melatonin did not enter the standard of care for treatment of the condition.

Yet the truth is, the study wasn't negative; it was inconclusive. It was underpowered. To be clear, the study doesn't imply that melatonin works—the observed results, though dramatic, are entirely consistent with a chance finding. The study simply did not have the power, the resolution, to detect a clinically meaningful effect. For me, even a 10 percent difference in outcome rates would be clinically meaningful—transformative, even. But the size of that clinically meaningful outcome was not large enough to be seen by the microscope of this particular study. A much larger study, a more powerful microscope, would be needed to detect that important but smaller effect size.

Given that, some argue that conducting an underpowered study is inherently unethical. If the study is underpowered to detect a clinically meaningful difference in outcomes, why even bother doing it? We are merely exposing individuals to the risk of a novel intervention without any expectation of scientific benefit. While I support this view in some cases, I believe there is a role for some small randomized trials—particularly to examine certain biochemical treatment effects and to help aid in the design of more definitive studies.

We've touched on the issues of statistically significant findings that don't matter and statistically insignificant findings that might matter, but doing so may have given you a sense that the only things that can go wrong with clinical trials are how the results are interpreted. If only.

Far from Perfect

I spend my lunch breaks alone, at my desk, with my office door (which is ordinarily open for students, trainees, and other investigators) closed. It's a little bit of time I carve out for myself—twenty minutes or so, if I can make it work. As I eat, I pull up YouTube and watch my dirty little secret: math videos. That I find these ruminations on the properties of prime numbers or the irrationality of the square root of 2 peaceful no doubt says something about my psyche that I have yet to uncover, but I think I just like the certainty of it. In math, you know if you got the right answer. You put in the work, you come up with an answer, and you can check the result—often in multiple ways. Math is a certain science.

Math coalesces to form physics, still fairly certain, as science goes, but not quite as perfect. The laws of physics dictate the laws of chemistry, and the laws of chemistry, the observations of biology. And where biology meets humanity is Medicine—imprecise, imperfect, and yet more relevant to our experience as humans than any other form of science.

While the randomized trial is really the pinnacle of medical science, the bar is not that high. The randomized trial never comes close to claiming certainty, like a mathematical proof does. It simply measures probability. *This* treatment is probably better than *that* treatment, the trial says, with the following amount of uncertainty. And I have spent enough time lauding the randomized trial paradigm that it behooves me to tell you how randomized trials can go wrong.

The simplest way to take a nicely designed randomized trial and botch it is to lose track of the people you randomized. The technical term here is "loss to follow-up," and though I'm sure you're thinking that this must be incredibly rare, it is far, far from it. People move and don't let the study team know where they went. People die unexpectedly. People stop returning phone calls or showing up for study visits. Nearly every randomized trial has *some* loss to follow-up, and a general rule is that somewhere south of 10 percent is "acceptable"—though there is a major caveat to that, which I'll get to in a moment.

The problem with loss to follow-up is that the whole point of the randomized trial is to figure out what happens to people who get treated with your new therapy versus the control group. If you lose track of 10 percent of them, you don't know what their outcomes were, and it makes it harder to make good inferences about the true effect of the drug.

It turns out that if people leave the study at random, which is to say in a manner unrelated to which randomization group they were in, it's not a big deal, though you increase the risk of having a "negative" study. Some RCT planners recruit extra participants to be sure that, even if they lose track of some, they'll still have adequate numbers to figure out whether the drug works or not.

Now that caveat: If the loss to follow-up *differs* between the treatment groups, you are in big trouble. Imagine that I invented a drug that I hope leads to weight loss. But there's an unfortunate side effect of this hypothetical blockbuster: mind-melting nausea. In that case, patients in the treatment arm might be lost to follow-up much more frequently than those in the placebo arm—after all, there's a good chance they'll want nothing more to do with my study after the first pill they take! That means, in the end, the only people I can analyze in the treatment group are the tough folks who could muscle through the terrible side effects. Needless to say, that is *not* a random group of people. In this way, differential loss to follow-up breaks the careful randomization that was the intent of the study all along.

The starkest example I've seen of this was a hugely ambitious but ultimately fatally flawed study that appeared in *Science* in 2014. Now, *Science* is a big deal. Along with *Cell* and *Nature*, it is one of the "big three" journals that basically every researcher aspires to publish a paper in. Careers are made with less. The study was incredibly interesting. In the 1970s, 122 disadvantaged kids were randomly assigned to receive a complex intervention that included early childhood education, nutritional support, and healthcare or to not receive those interventions. The study in *Science* looked at their cardiovascular health outcomes thirty years later. The conclusion? The multipronged intervention significantly reduced the risk of cardiovascular disease and diabetes.

This is the kind of study that gets massive attention, and rightly so. First of all, thirty years of follow-up for any study is remarkable—for a randomized trial, it is nearly unheard of. Second, it provides concrete, actionable information that could be critical for the public health. And, third, it has a good message: Investments in children pay dividends in adulthood.

But there was a problem. Thirty years is a long time, and many of those kids were lost to follow-up. In fact, 31 percent of boys in the treatment group were lost to follow-up, and 48 percent in the control group. We don't know what any of their cardiovascular outcomes were. This is a huge amount of loss to follow-up and, importantly, differential loss to follow-up, which means that the group being analyzed in their thirties no longer benefits from the secret sauce of randomization. With randomization broken, the ability to assess the causal link between those childhood interventions and adult health is also broken.

In my heart, I actually do believe that early childhood intervention would improve long-term health in adults. But this study can't show us that, and we have to be honest about the inadequacy of data even when the data supports our previously held beliefs.

Loss to follow-up is about missing outcomes. But the other major way randomized trials can mess up is by failing to isolate the intervention. In other words, the conclusions are wrong from the start. Let's say I want to test whether my new pill to reduce depression is effective. To ensure the integrity of my study, I decide to call the people randomized to take the pill every day to make sure they aren't having side effects that could be troubling. Since the placebo is just a sugar pill, I don't bother calling the participants assigned to a placebo.

As you might expect, I would be hard-pressed to conclude that any difference in depression scores I observe between the groups is due to my pill. Couldn't it be because they got a phone call from a concerned doctor every day, checking in on them? This is formally known as "co-intervention bias," and it is a highly common way randomized trials of alternative therapies go wrong.

One of the things that makes acupuncture or therapeutic touch or reflexology or aromatherapy so difficult to study is that the interventions come in a "package." Acupuncture isn't delivered in a capsule you swallow just after you brush your teeth but before you brush your hair. It is delivered in a quiet room, with appropriate lighting, soothing music, and a gentle, soft-spoken practitioner telling you everything will be okay. It is incredibly difficult to mimic *every* aspect of this experience except the acupuncture.

The best studies use "sham" acupuncture—whereby the entire experience is replicated but the needles are placed in areas that do not correspond to the underlying acupuncture philosophy. With rare exceptions, the vast majority of acupuncture versus sham acupuncture randomized trials show the same thing: Sham acupuncture works about as well as real acupuncture, and both work way better than nothing. In other words, maybe the lighting and companionship of the acupuncture practitioner (and the meditative quality of the experience) drive the benefit patients experience far more than the needles entering the skin at certain energy points.

Obsessed with Imperfection

To be sure, there are plenty of other ways randomized trials can go wrong. Researchers can use the wrong placebo or no placebo at all. They can fail to blind their analyses. They can even commit outright fraud (see chapter 9). But I'm obsessed with randomized trials because I am a pragmatist, and I realize that, with all their flaws and potential for error, they are still the best tool we have to really know which medicines work.

Given all these caveats, I wouldn't blame you if you feel like you don't know which studies to believe and which studies not to believe. If "significant" doesn't mean significant and "negative" doesn't mean negative, is modern medical science just a form of modern art, something each individual is free to interpret as they wish? Fortunately not. But it does mean that we need to go beyond the headlines that report trial results. We need to look at those results ourselves.

This may seem daunting, but it really takes just a few simple steps. First, identify what the exposure of interest was and what the outcome of interest was (i.e., what were they testing, and what effect did they look at?). Second, look at the rates of that outcome in the treatment and control groups. Third, check the p-value to evaluate the statistical significance of a study. Fourth, check your gut to evaluate the clinical significance of a study.

For example, you come across a headline that reads RANDOMIZED TRIAL SHOWS A SIGNIFICANT BENEFIT OF YOGA IN PEOPLE WITH HEART DISEASE. You can determine the exposure of interest (yoga) from the headline. But what is the outcome? What is this "benefit" the headline speaks of? Reading the article, you find out that the benefit is measured in terms of lowering stress levels as measured by a standard survey. Reading further, you find that the group assigned to do yoga had a one-point reduction in stress (on a scale of 100) compared to a three-point *increase* among those assigned to no yoga. The difference is statistically significant at a p-value of 0.04. You decide that the study is interesting but, given the small effect size, not clinically meaningful to you. You may decide to keep your eye out for future yoga-based randomized trials, with the hope that new techniques might have an even stronger effect on stress levels. After all, interventions that truly work should be easy to replicate.

In chapter 3, we discussed "one thing" medicine—the false idea that by changing some simple aspect of your life you can be protected from the chaotic winds of an unfeeling universe. We can say the same thing about randomized trials. No one randomized trial will show you the absolute truth about a given medication. But it will move you *toward* truth. And every successful trial will get you closer and closer to true understanding. Every randomized trial has a measure of uncertainty, but uncertainty decreases as more randomized trials of an intervention are performed.

The big decision to make with your doctor, then, is when to pull the trigger. When is the evidence good enough? This is where we can build trust together: by realizing that, while randomized trials are rather cold and indifferent, the application of knowledge gained from randomized trials is intensely personal. You can use the knowledge you've gained here—about the secret sauce of randomized trials and the fact that sometimes that sauce goes bad—to have a discussion with your doctor about where the data really lies. Are we flying blind, like I was in the early days of the coronavirus pandemic? Or have enough studies been done that we can speak with near certainty? For some of us, one promising trial may be enough to try a new therapy, particularly if there are limited alternatives. Others will feel that my favorite line—"More data is needed"—serves as a mantra to defend against false hope. And that's okay.

In this framework, it becomes clear who the real heroes of medical science are: the participants in these studies. Those individuals who agree to the cold, indifferent treatment allocations of a coin flip. They buy our increasing knowledge with their bodies, often to no personal benefit. It is an act of altruism that I wish our society and governments would reward more directly. For without those volunteers who take it upon themselves to be the subject of a rigorous experiment—but an experiment nonetheless—there would be no medical progress, no new knowledge, no breakthroughs, and no cures.

CHAPTER 6

Good Medicine May Not Be Good for You

MY CHILDREN ARE terrified of mushrooms, and it's entirely my fault. As soon as they were old enough to notice those fungal fruiting bodies, they were quickly instructed never to touch them, never to pick them, and, for all that is holy, never to eat them. At this point, they essentially consider wild mushrooms to be tiny immobile agents of death, and give them a particularly wide berth. While this may seem like an extreme reaction to have instilled in my children, I'd rather they play it safe. After all, I've seen firsthand mushrooms nearly kill an entire family, in a situation where life and death came down to a game of percentages.

The James family lived in a modest three-bedroom house in the suburbs of Philadelphia. Mr. and Mrs. James worked in the public school system where their three children (two girls and a boy) attended school. Their oldest daughter was eighteen, preparing to head off to college in the fall. They were a typical American family, until one day, in the spring of 2008, they all ended up in the

Hospital of the University of Pennsylvania at the same time. The reason? Mushrooms.

The family had had some experience harvesting wild mushrooms in the past, and when a crop of benign-appearing white-capped medium-round mushrooms appeared in their backyard, they picked them, chopped them into a spinach salad, and made an early meal of them. About six hours later, they started developing their first symptoms—abdominal pain, nausea, and diarrhea. By twelve hours, the nausea and vomiting had gotten bad enough that they sought care at the hospital. By twenty-four hours, all five members of the family were in acute liver failure. The two parents were on the adult ward in guarded condition. The two younger children were in the pediatric hospital next door. The condition of the eighteen-year-old girl had deteriorated the most rapidly; she was in critical condition in the ICU.

The vast majority of mushroom toxicity cases are cases of mistaken identity. Amateur mushroom hunters misidentify a toxic mushroom as edible. The most common toxin in toxic mushrooms, amatoxin, is heat-stable; cooking it does no good. When amatoxin is ingested, it is rapidly absorbed through the intestinal lining, where it is transported to the liver and absorbed by liver cells. This is where amatoxin does its damage.

All cells depend on the constant generation of new cellular proteins to replace those that have worn out, and to allow the cell to adapt to new conditions. The instructions for making these proteins are encoded in the sequence of DNA, which is protected in the nucleus of our cells like a book of blueprints kept at the central office of an architecture firm. Those blueprints are translated into messages by an enzyme called RNA polymerase and sent out of the nucleus into the cytoplasm, the construction site for new proteins, where they are pieced together from amino acids.

Amatoxin blocks RNA polymerase. No messages leave the

nucleus to get to the cytoplasm when amatoxin is present. The construction site stands ready to produce necessary proteins, but no instructions arrive. Without a steady influx of new proteins, cells die. The medical term is "fulminant hepatic failure," an abrupt decline or cessation of liver function in the face of extensive liver damage. It can be seen in Tylenol overdose, from exposure to certain chemicals, from a lack of blood flow to the liver in the setting of shock, and, infrequently, from certain mushrooms.

Within seventy-two hours of ingestion, the eldest daughter of the James family, Megan, had become delirious. Her liver, an organ that removes metabolic byproducts from the blood, had failed, and those byproducts had caused confusion and stupor. She had started to bleed excessively from the sites where IVs entered her body, the liver being the primary organ that produces the factors that cause blood to clot. We were transfusing red blood cells, platelets, and coagulation factors around the clock to try to compensate for the lack of liver function. Our efforts were a poor imitation of the functioning of a healthy liver, but we were keeping her alive.

By day 6, the other family members were out of the woods, recovering on their own after some touch-and-go moments. Megan was still in the dark forest. Her mom and dad were able to visit her in the ICU, though by this point she was unresponsive, requiring a breathing machine, and deeply jaundiced.

Doctors from around the hospital had rallied to this case, which required input from multiple subspecialties, and I was privileged to sit in on a family meeting where the lead physicians discussed next steps with her parents. At the table were the ICU attending physician, the nurse who had taken care of Megan more than anyone else, the unit pharmacist, and the liver transplant team.

The parents had a decision to make. One cannot live without a liver, and Megan had very limited liver function according to her most recent lab work. She could be listed for a liver transplant and, given her critical illness, might be able to receive one within a few days. The other option was to wait and hope that the natural regenerative ability of the liver would kick in on its own. But if she got worse while waiting, her chances of surviving a liver transplant operation would drop to near zero.

She was on a razor's edge of risk. A quick liver transplant might save her life but would change it forever. She would require immunosuppressive medications daily for the rest of her life, and would be on a clock that no eighteen-year-old should be on: Liver transplants have a shelf life ranging from five to twenty years; she would likely need another in the future. Or we could do nothing, in which case she would either die or recover and have a chance of living an essentially normal life. It was an all-or-nothing play.

I knew what the data showed. Given how sick she was, the chance that she would recover without a liver transplant was small. I knew that, on average, the rational choice here was to do the transplant. Yes, your daughter would have a lifetime of medical issues related to being a patient with a transplanted liver, but you'd still have a daughter.

Yet the average choice is not always the right choice. And it wasn't for the James family. Maybe they saw something in Megan we didn't. Maybe they just looked at the decision through the optimistic lens of parents who want the world for their daughter. Maybe they were engaging in motivated reasoning and convincing themselves that she wasn't as sick as she actually was. But they refused the listing for transplant. We could reconsider in a few days, they said, and we agreed. Privately, I worried that in a few days it would be a moot point.

By day 8, we started to see some improvements. The blood clotting factors started to rise, her jaundice seemed to lessen slightly (first in the whites of her eyes, and later in her skin). By day 10, she was responding to simple commands. By day 12, she was breathing on her own and talking to her parents. As close to death as she had been, she required a few weeks of rehab to get her walking steadily and eating well again, but within a month it was clear that the James family had played the odds and won. Megan had recovered.

PATIENTS OFTEN BELIEVE that there is a single, "right" answer to their medical problem. And while we discussed in chapter 3 how we can all be led astray by "one thing" medicine, there is still the implicit belief that when we have to make a choice between options, there is a right one and a wrong one. Really, all of medical science is about trying to figure out what the right choice is in a given clinical scenario. Even studies that compare a new drug to a sugar pill are implicitly studying a choice: Is taking this drug better than doing nothing? (Often, the answer is no, and the drug does not make it to pharmacy shelves.)

But medical studies look at groups of people, and you are not a group. You are an individual. Making good medical decisions doesn't *always* mean doing what has the best chance of success, on average, because you are not average. No one is. Teasing apart the individual benefit of a medical intervention from the population-level benefit is complicated and something most doctors don't fully understand. But to be able to really talk with your doctor, and to explain why you might be making a choice that seems "wrong" to him or her, you need to understand the difference between the "average" person and you.

On Individual Risk

When making decisions, we often frame our thinking in the language of risk versus benefit. Should you invest in that new company?

Should you eat that extra slice of cake? Should you have an affair? Rob a bank? Ask your boss about a promotion? Take a new medication? The risk-benefit paradigm is comforting. It makes decision-making feel rational, and divorces it from emotions—and, in some cases, ethics. Most of my patients frame their medical decisions in this way, trying to estimate their risk of having a side effect of a medication versus the potential benefit of taking the medication. And though we don't always agree on the particulars of those estimates, their rationale is solid.

But what is "risk"? We all have an intuitive understanding of the term, but that understanding breaks down when it is probed a bit more deeply. When I teach statistics to medical trainees, I start pretty simply. Let's say that a medical procedure has a 10 percent risk of death. What does that mean? Well, the straightforward answer is that it means that if one hundred people get the procedure, we can expect ten to die. Or if one thousand people get it, we can expect one hundred to die. Risk conveys expectations about the behavior of a group.

But what does risk mean to the individual? To the person undergoing the procedure? If I tell you that *you* have a 10 percent chance of dying during the procedure, how do you interpret that number? You will either die, or you will not. There is no such thing as 10 percent dead. At an individual level, we can conceptualize risk only as it relates to something else. A 10 percent risk of death seems incredibly high, because there are few things in life that we have ever faced in which one would expect one of every ten people to die.

Moreover, the risk of any decision needs to be conceptualized against the background risk of *not* making the decision. If I told you that the procedure with the 10 percent mortality rate was for a condition that was universally fatal if untreated, a mere 10 percent mortality rate would seem to be a miraculous success. What this all boils down to is that the risk to *everyone* and the risk to *you* are different conceptually, and the mathematics of risk don't capture this. To illustrate this, let's go to Vegas.

The Slot Machine Problem

We walk past the fountains of the Bellagio and enter the lobby, taking note of the two thousand blown-glass flowers that dot the ceiling. But we are not here to gawk; we are here to make money. Because tonight, for once, the odds are in our favor. You see, one of my data science interns, in an effort to please his cantankerous boss, has hacked into the servers controlling the slot machines and adjusted their payouts to our benefit. Two machines were successfully hacked. We just need to decide which one to play.

Hacked slot machine #1 costs $1 per pull on the lever. The payout for a jackpot is $200. And thanks to my data science intern, we will hit that jackpot, *on average*, one out of every one hundred times. These are great odds, as you can see—chances are that by spending $100, we will win $200. It's not guaranteed, of course, but it's a hell of a lot better than the normal house odds.

Hacked slot machine #2 is in the high-roller area. It costs $1,000 per pull on the lever. Like slot machine #1, the data science intern has hacked it to pay out, *on average*, one out of every one hundred plays. But this machine has an even better jackpot: $300,000. That's right. We have a one in one hundred chance for a jackpot on both machines, but the jackpot for machine #1 is two hundred times the price to play, whereas it is three hundred times the price to play for machine #2.

Which machine is the better choice? Numerically, it's clear—you will make more money playing machine #2. But are you going to play it? I sure wouldn't. Because I don't have the kind of money to play machine #2. Sure, I will come out on top in the end, assuming I keep playing again and again and again, but it will cost me $1,000 each time I pull that handle, so I simply can't play that much. If I need to play one hundred times to be pretty sure I'll win, I may need to spend $100,000 before I succeed. And while there's a chance I might win on my first try, there's also no *guarantee* that I would win by my one hundredth try. I am going to stick with the $1-per-pull machine.

The slot machine problem illustrates how the most rational choice in a situation can change depending on how many times we get to make that choice. If the rules of the casino were that you get only one pull on a slot machine before you are kicked out forever, you'd pull the $1 slot—chances are you'd lose (you win only one out of one hundred times), but at least you've lost only a dollar. Conversely, if you could afford to play the $1,000 machine a few hundred times, you definitely would—you'd come out further ahead than if you played the $1 machine instead. All of these choices are perfectly rational.

The thing about slot machines is that they are random but, at the same time, perfectly predictable. While we don't know how any given pull of the handle will play out, we can know with precision how one hundred, or one thousand, or ten thousand pulls will play out. In a way, it's like that procedure with the 10 percent mortality rate. That estimate may be perfectly accurate for a group of people undergoing the procedure, but it doesn't really tell us what will happen with *that particular patient over there.*

As doctors, we get a lot of pulls on the slot machine of health; we can prescribe the same medication to an entire panel of patients, knowing that, on average, some of them will benefit. But that may be cold comfort to those patients suffering through medication side effects and not deriving any benefit. In other words, the best, most rational choice a doctor can make may not always be the same as the best, most rational choice a patient can make. Fortunately, there is a way doctors and patients can speak the same language on this issue.

The Biggest Secret in Medicine

To talk to your doctor about personal risk (instead of population risk) requires you both to understand what I think is the biggest secret in Medicine. It is a secret not because of some vast conspiracy, but because it is poorly understood. I have a particular disdain for conspiracy theories that require whole professions to be in on the grift, and yet I hear all the time that we doctors are withholding miraculous cures from the public. Cures for cancer, for diabetes, for weight loss, for aging—all of these are purported not only to exist, but to be widely known by physicians like me, who actively keep them hidden. The typical story goes that we are being silenced by Big Pharma or taking kickbacks, and the whole system would come crashing down if word got out that a bit of turmeric is all you need to cure acute lymphoblastic leukemia.

This is, of course, ridiculous. First, because most doctors are not psychopaths. But even more practically, keeping secrets is really hard. You probably know from experience that the more people who know a secret, the more likely it is that that secret will come out. Couple that with the fact that, as I mentioned earlier, most medical researchers have a deep and abiding interest in becoming famous for curing some terrible disease, and you will realize there is really no way a miracle cure would be kept behind closed doors.

But there is one secret that the entire medical establishment seems to keep, likely because most doctors aren't thinking deeply about it. If you understand it, you will be way ahead of the curve of most patients and many doctors. Stated simply, it is this: Chances are, the medication you are taking isn't going to help you.

Now, before you run off and shout that this physician-scientist thinks that all medicine is nonsense, let me draw attention to something very specific in the way I wrote that secret. *Chances are*, the medication you are taking isn't going to help you. But it might. Scientific studies show us with a high degree of certainty that a medication will help *some* people with a given condition. The problem is, we have almost no ability to know specifically who among the people with that condition the medication will help and who it won't.

Let me start with a practical example: How low should your blood pressure be? For a long time, high blood pressure (hypertension) was defined as a systolic blood pressure (that's the top number) greater than 140. Normal blood pressure was considered 120 or less. So what to do with individuals in the 120 to 140 range, the blood-pressure gray zone? It wasn't clear. Some docs were more aggressive, and would add medications to get the numbers lower. Some would advise lifestyle modifications, like reducing salt intake and getting more exercise.

But in 2015, the results of a hotly awaited clinical trial called the Systolic Blood Pressure Intervention Trial were released. SPRINT had enrolled almost ten thousand people and split them, at random, into two groups. One had the standard blood-pressure target of 140. One had an "intensive" blood-pressure target of 120. After just over three years, the study, which was supposed to run for at least five years, was stopped. The reason? There were substantially fewer deaths in the intensive control group. In fact, the results were so compelling that an external monitoring group concluded it would be unethical to leave people in the standard-treatment group any longer. Trial over—everyone gets intensive therapy from here on out.

SPRINT was a well-designed, well-executed, high-quality study that answered a clinically important research question. The results made it clear that targeting 120, instead of 140, would save lives. This is something we doctors really like to do, and the study led to widespread changes in hypertension treatment guidelines.

But the splashy headlines don't capture the whole story—at least, not for individual patients. The rate of death was reduced in the intensive-treatment arm, but, as you might imagine, not that many people died overall. Death is, fortunately, a rare event, even among people with high blood pressure. There were 155 deaths in the intensive-treatment arm, or 3.3 percent of participants, and 210 deaths in the standard-therapy arm, or 4.5 percent of participants. Sure, 3.3 percent is better than 4.5 percent, but, realistically, most people in both arms made it out just fine. What does this mean for you, the individual patient? Not the same thing it means for a physician.

Put yourself in the shoes of a physician treating patients with high blood pressure. Let's say you have a panel of one thousand such people, and you'll be treating them for just over three years. If you adopt the SPRINT practice of intensive blood-pressure treatment, you can expect thirty-three of your one thousand patients to die. If you stick to the older practice of standard blood-pressure treatment, forty-five of your one thousand patients will die. That means that your choice, intensive versus conservative treatment, will save twelve lives.

But what is the cost of that choice? To save those twelve lives, you need to treat one thousand patients with additional medications to get their blood pressure down to below 120. We can divide that one thousand by twelve to arrive at a statistic called the "number needed to treat," or NNT—the number of people you need to treat to save one life. In this case, the NNT is 83.

Is it worth treating eighty-three people with a certain practice to save one life? The answer, like so many answers in Medicine is,

"It depends." If the intervention is cheap, without side effects, and widely available (like a vitamin C pill or a daily walk), then, sure—go for it. If the intervention is highly toxic and expensive (like chemotherapy), it may not be worth it. In the case of blood pressure, the burden and cost of an additional blood-pressure pill is not too bad, but reasonable minds may differ here.

Let's circle back to that NNT of 83. There are 108 million people in the United States with high blood pressure right now. If you were to treat all of them with intensive blood pressure lowering à la SPRINT, you'd save more than a million lives. It's a public-health slam dunk.

But what about you, the *individual* patient? Let's say you have high blood pressure, controlled with medication to a level of 140. Your doctor says something to the effect of "I'd like to get your blood pressure down to one-twenty, in accordance with this slam-dunk clinical trial. It will require taking one extra pill a day." You ask what the chances are that the extra pill will save your life. "Well, one in eighty-three," your doctor says.

And that's the central problem. A physician's duty to society, in this case, is clear: Treat everyone intensively to save lives. But our duty to our patient is not as clear. Is it ethical to saddle you with an additional pill every day (including the costs and side effects associated with taking that pill) when there is an eighty-two out of eighty-three (99 percent) chance it won't make a difference in whether you die or not? The problem is that our best tool to assess the true efficacy of an intervention—the randomized trial—tells us about the population effects of a treatment. We may know quite precisely that if we treat everyone with a major heart attack with clot-busting drugs, we will save one out of every one hundred patients, but we have no reliable way to know *which* one patient that will be. In the absence of that information, the only rational choice as a physician is to treat everyone. We have the luxury of multiple pulls on the slot machine, but you, the patient, get only that one pull—yes or no—take the drug or not. Do you feel lucky?

This is why it is possible for you and your doctor to disagree on a course of treatment, and for both of you to be acting entirely rationally. It's not that one of you is right while the other is wrong—you are both right, from the perspective with which you are each approaching the question. Doctors tend to approach these issues by thinking of the group benefit, as we have multiple patients to treat and the best data we have examines the effect of an intervention at the population level. Patients, appropriately, are thinking of the individual benefit and risk.

Doctors can't force patients to do anything, but key to building and maintaining trust is to come to a decision that both parties understand (even if they don't fully agree with it). The language of the "number needed to treat" can be an incredibly useful tool to discuss the real risks and benefits of a medication choice. It allows you to say "Yes, I understand that this medication may save my life. I also understand that chances are I'll be fine without it, and given my risk tolerance, I am willing to roll the dice on this." Your doctor may not agree, but they are not the one who has to pay for or take the medication. In the end, it is your decision. You may decide even a 1 percent chance of a better outcome is worth it. Our primary goal as physicians is to make sure you are making that decision with the real information. The NNT helps with that.

By the way, you may think I picked the SPRINT study because it has a particularly large NNT, but NNTs in the tens or even hundreds are far from uncommon. A wonderful website, TheNNT.com, catalogs some of them. A smattering:

- Taking aspirin after a major heart attack: NNT 42 to save one extra life
- Antibiotics for sinus infections: NNT 15 to cure one extra patient

- Dexamethasone for hospitalized patients with COVID-19: NNT 36 to save one extra life
- CT scans for smokers to detect lung cancer early: NNT 217 to save one extra life
- Vitamins to prevent heart disease, stroke, or death: NNT infinity. (In other words, the major studies show no effect. But you knew that already.)

I mentioned earlier that the reason it's a big secret that the medication you're taking may not help you is because doctors don't really think about it. In part, this is because we are not trained to think in these particular terms. We are taught to learn what intervention works for a particular disease, and we internalize that as the standard of care. And evidence suggests we vastly overestimate the efficacy of these interventions.

A study that appeared in *JAMA Network Open* in 2021 brought this fact into stark relief. Five hundred forty-two clinicians were presented with a series of case scenarios and asked about therapeutic options. One, conveniently enough, presented a patient with mild hypertension. The clinician was asked how likely the patient was to have a cardiovascular event (stroke or heart attack) within the next ten years. The "true" answer, based on epidemiologic data, is around 3 to 12 percent. The average answer among clinicians was 10 percent. So far, so good—clinicians had a good sense of prognosis.

But the study then asked how likely the prescription of a blood-pressure-lowering drug would be to *prevent* a cardiovascular event. The average answer was 30 percent. Clinicians felt there was a one in three chance that the drug they prescribed would stave off a heart attack or stroke. In reality, as you know, the real chance is somewhere closer to 1 or 2 percent.

This doesn't mean the drug is bad. Many of us would do

something that has a one in one hundred chance of saving our lives. Wearing your seat belt, for example, has an NNT of around 25,000 (after all, most people won't die in car accidents whether they have a seat belt on or not), but we still do it, because it is cheap, easy, and basically without side effects.

The *JAMA Network Open* study shows that doctors, while accurate at predicting the likely outcomes of disease, overestimate the benefits of medications. This may be a subconscious effort to avoid despair at our weak anti-disease arsenal, but I think it has more to do with that population mindset again. And I believe that a major culprit, in fact, is the way results are reported in medical studies.

Relative Risk Is for Populations, Absolute Risk Is for Individuals

You may think there is only one way to present the results of a big medical study: showing the percent of patients with the outcome of interest in each treatment group. If only. While there are myriad approaches to conveying the truth of the underlying data, they break down into two big categories: relative risks and absolute risks. The former is good for physicians who like to think about populations, and the latter is good for patients who prefer to think about individuals. Of course, the former gets a *lot* more airtime. Here's how they work.

Let's say you have a disease with a 1 percent mortality rate—something like COVID-19. Given one hundred people, one will die and ninety-nine will recover. You test a drug in a large study to see if it can help, giving a random group of afflicted patients a placebo and another random group the new drug. The death rate in the placebo group is 1 percent, as expected. In the new drug group, the death rate is 0.5 percent. You can breathlessly report that your drug cut the death rate in *half.* This is a 50 percent reduction in deaths. Huge.

What you have just described is the *relative* risk reduction—and it is really important at a population level. If one hundred thousand people per year die of this disease and you give everyone with the disease your new drug, only fifty thousand people will die. You cut the number of deaths in half. Amazing. But the *absolute* risk reduction is not 50 percent; it's just 0.5 percent—that's how much you lowered the death rate. You converted a disease with a 1 percent death rate to one with a 0.5 percent death rate. And it's the absolute risk reduction that dictates the NNT.

If we imagine one thousand people with the disease, we would expect ten to die under ordinary circumstances. Now just five die. So we treated one thousand people to save five lives. NNT=200. The chance that this drug will save any *individual* patient is really quite small. After all, the vast majority of patients survived this disease even before we invented our blockbuster therapeutic. When it comes to the individual, absolute risk matters, not relative risk.

This difference explains why the doctors in the *JAMA Network Open* study were so bad at estimating whether their drug would help their patient—they were thinking in population terms, not individual terms. In our hypothetical example, their intuition would suggest that the drug they are prescribing has a huge chance of helping that patient (after all, it cuts the death rate in half), but they forget that the death rate is already small. For that reason, the chance that the drug will save the patient's life is also small (albeit larger than doing nothing!).

One trick question I ask my students when we are reviewing a new medical study of a promising drug goes like this: "So this study randomized some patients to the drug and some to a placebo, and significantly fewer died in the drug arm. What do you think the number one predictor of survival in this study is?" The answer they always give is "Whether you were lucky enough to be in the drug group." I then walk through the data to show what the answer actually is. It's never which group the participant got randomized into.

In the SPRINT study, the number one predictor of survival was younger age, followed by a lower heart-risk score, followed by better kidney function, nonsmoking, lower blood pressure, and a variety of other risk factors. Of thirty-one potential factors in the publicly available SPRINT data set, getting randomized to the intervention arm was the twentieth-best predictor of survival.

What makes medications special is not the fact that they are so incredibly good at saving lives, but that they are easy to use. You can't change your age, and quitting smoking is tough, but popping a pill is a straightforward action that doesn't demand too much change in your usual habits. The sobering truth is that drugs operate on the margins—they have an effect, but they aren't as impactful as we like to think they are.

Wouldn't it be nice, though, to be able to know that you were the one out of one hundred people whose life *would* be saved by intensive blood-pressure control, or one of the ninety-nine where it wouldn't matter? Is there a way for doctors to make the best choices with their patients while still preserving the benefit to society at large? There may be, but almost no one in Medicine is using it.

Uplift Modeling

Have you ever gotten an ad that pushes you over the edge toward buying a product? Like you'd been thinking about purchasing that new phone or car or pizza oven but hadn't quite gotten to the point where you were ready to buy. And then you got some ad, maybe with a discount code, and you decided to pull the trigger? Of course, this has happened to all of us. And this is exactly what marketers want when they are making these ads. The people they really care about are those already on the edge, close to buying the product—they just need a little push.

Imagine you are the head of marketing for a diaper company, and sales are down. Your hunch is that a coupon campaign might really goose the market, but you want to be rigorous about it. You decide to split your mailing list at random and send half of your customers a one-dollar-off coupon, while the others will act as a control group. After a few months, you track those purchases and determine that the customers who received a coupon were 10 percent more likely to buy diapers. Success! You tout your rigorous study to the higher-ups at the company, expecting their plaudits and perhaps that promotion you've been angling for. But that's not quite the reception you get.

"Wait a second," they say. "How much did we spend on postage sending coupons to people who didn't buy diapers at all? What about the people who were going to buy diapers anyway? We basically gave them free money!" What the fat cats in the C-suite want is for you to find a way to send coupons only to those people (known in the marketing biz as "persuadables") who will not buy a diaper *unless* they have a coupon. The people on the edge. You should not waste your time or money on anyone else.

With a moment's thought, you can see how this example parallels the central dilemma of Medicine. We do a trial and determine that A is better than B, but applying A to every single patient is wasteful (like sending a coupon to every single person on the mailing list). The ideal would be to give the treatment *just* to those who would benefit from it—to send the coupon *just* to those for whom the coupon will change their buying behavior.

Marketers have largely figured this out with a series of mathematical models called "uplift algorithms." These statistical tools look at data (as in the coupon experiment) and figure out the characteristics of those customers for whom a coupon really moves the needle. Marketers can then send those coupons to just the individuals for whom the coupons are likely to be highly effective.

You may not be in the diaper-buying game anymore (I, thankfully, am not), but you are almost certainly part of an uplift algorithm. Those ads you see on Amazon that seem perfectly targeted to you? That's an uplift algorithm. The sponsored posts on Facebook? Uplift. Even political campaigns are in on the game—testing various versions of their emails begging for donations, and sending you the one that, according to their uplift model, is most likely to get you, individually, to open your wallet.

Uplift algorithms seem ideal to break the conflict between a doctor's obligation to a patient and to society, because they would allow us to take a medication that benefits one out of one hundred people and figure out who that one person is. Or at least get closer to figuring it out, such as identifying ten people, with some confidence that the one is among them. The problem is, these algorithms aren't being formally tested in Medicine. Physicians and researchers seem content to stick with the old paradigm: treating everyone the same despite realizing that that policy will benefit only a subset.

I think this is primarily due to a lack of familiarity with these advanced models rather than a nefarious conspiracy, but it is worth mentioning that the number one group to lose if uplift modeling were widely adopted (and shown to be effective) is pharma. Treating *fewer* people but achieving the same number of lives saved is, frankly, bad for the drug companies' bottom line. That means that you'd be hard-pressed to find a pharmaceutical company willing to fund research studies looking into uplift modeling, which means we may be left with just the government (via the National Institutes of Health) to provide support for this type of research.

The utopian day when your doctor can tell you with 100 percent certainty that a medication will benefit you as an individual is, therefore, quite a ways off. So what are you to do in the meantime? With every medical decision, you have a simple question to ask: Is it worth it?

So *Is* It Worth It?

At this point, you might be feeling quite discouraged. If you are the type who really doesn't like taking pills, you might even be reading this chapter as permission to stop taking medication altogether. If that is the case, check your motivated-reasoning sensor again. The fact is, even if a medication has a low chance of helping you, it absolutely could still be worth taking. You wear your seat belt, despite the fact that there is only a one in twenty-five thousand chance that it will save your life someday, because it *may* save your life someday. You never know when you might be "the one" who would die or have some other adverse consequence of forgoing an intervention. So the interpretation of the "number needed to treat" needs to be done in the context of the costs (both monetary and in terms of side effects) of the medication or intervention.

There are also a few caveats to keep in mind before you throw all your pills down the drain. First, consider the fact that every medication has its own NNT. Though the chance that you will be the one person out of ten or fifty or one hundred to benefit from a specific medication may be low, each medication you take is a new chance—a new pull on the slot machine handle. Your medication list is, in essence, a hedged health portfolio. We aren't sure which of these pills will really help, but all together the chance that one might make that life-or-death difference is higher.

Also realize that NNT numbers tend to be a bit conservative, due to the fact that the studies that generate them are conducted over a specific period of time. SPRINT lasted three years, so that

NNT of 83 suggests that aggressive treatment of eighty-three people with high blood pressure saves one life *over three years of treatment.* For many chronic conditions, treatments last way longer than that, increasing the chance that your life will be saved at some point.

Additionally, note that the NNT doesn't just apply to medications. *Any* intervention—a diet, an exercise, a lifestyle choice—can be examined in the same light, and you will be disappointed in the NNTs for *all* of them. In chapter 3, I wrote about the impact of switching from a meat-eating to a vegetarian diet: NNT 5,500 to prevent one case of colorectal cancer.

Finally, realize that just as NNTs show that many patients won't experience direct benefit from a drug, the converse is also true; most patients will not experience side effects from a drug either. We can quantify side effects in terms of the "number needed to harm," or NNH, which looks at how many people you treat per one additional side effect (which could range from headache to nausea to allergic reaction, and so on). NNHs are quite a bit higher than you might expect, because side effects are common even for placebos! We are primed to look for any weird ache or pain after taking a pill, and that hypervigilance is why you see so many side effects reported from medications.

In actuality, most people will not get a side effect as a result of a new medication. For example, a side effect of intensive blood-pressure control could be fainting, and indeed there were more fainting episodes in the intensive-treatment arm of SPRINT than in the standard-treatment arm. But the "number needed to harm" was around 100. In other words, there is a similar chance that intensive blood-pressure control will save your life as there is that it will make you faint. You and your doctor can discuss whether the trade-off is worth it.

The NNT reinforces why a "one thing" philosophy leads us astray in Medicine. Because the chance that any one thing will be *your* one thing is often quite small, you need to do multiple things, hedge your bets, to lead a truly healthy life. We are quite right to roll our eyes at the latest diet fad, "superfood," or blockbuster drug. But dismissing all of them is the wrong move as well. We need to create a health portfolio that we stick to—a series of healthy behaviors, including taking medications that have a chance of working—because, all in all, that's the best way to optimize our chances at living longer and better. The best person to help you create that portfolio is your doctor, and you now have a shared vocabulary that can help you weigh the risks and benefits of your decisions.

Creating a Health Portfolio with Your Doctor

To be honest, if your doctor suggests starting a new medication and you ask what the NNT is, your doctor is unlikely to give you the exact answer. The NNT, fundamentally, is a bit of a wonky statistical concept that, while taught in medical school, is not frequently discussed in practice. Plus, as the 2021 *JAMA Network Open* article suggests, doctors tend to believe a medication will be more helpful than it is actually likely to be. So how does the NNT vocabulary allow you to build trust with your physician?

Most importantly, it means you can acknowledge the uncertainty around any treatment choice. You and your doctor can agree that there are no guarantees in Medicine, and that the real goal is to optimize your chances at leading a fulfilling life. Some choices will improve your chances a small amount, some a larger amount. And you can ask your questions in those terms.

Even without precise NNT numbers, you can use the concept of the NNT to compare two treatment options, for example. And you'll be able to acknowledge that even a small chance of benefit might be worth it, provided the costs are not too high. You can talk openly about those costs, what side effects to expect, and how likely you are to experience them.

This language will also help when you make a decision *not* to take a medication. Patients have the absolute right to refuse any medical care (provided they are of sound mind), and yet I admit that when I make a recommendation to a patient and they refuse it (or, worse, tell me they'll do it and then don't), it stings a bit—it chips away at the trust bond. But "I don't want to" and "I understand this has a chance of helping, but to me the risks outweigh the benefit" are worlds apart. The latter maintains our relationship and keeps us open to more discussion in the future.

In the end, I don't want my patients to do whatever I say. I want them to work with me, think through the issues, and make fully informed choices we both understand. To do that, we need to acknowledge that the right choice for *everyone* and the right choice for *you* might not be the same.

CHAPTER 7

No Such Thing as Incurable

EIGHT FOURTH-YEAR MEDICAL students were sitting in an apartment drinking beers and decompressing from what had been a hard week all around. It was 2006, nine months before graduation, and our tight-knit group had started our elective rotations, separating from one another for the first time in three years as we differentiated into proto surgeons, emergency room docs, pediatricians, and psychiatrists. Like all med students on their days off, we were sharing war stories—how early we had to wake up, how long we were in the hospital, the procedures we had done, the interesting cases we had seen, the cruelties and kindnesses of our supervising residents and attendings. Eventually, by beer three or four, we were discussing the fear that comes with being a physician. The fear of the magnitude of the job, but also the very simple fear that anyone coming face-to-face with severe disease every day would have: the fear of getting sick.

We talked about our fear of dementia, first—a visceral terror for young people whose life plans revolve very much around using their brains for as long as possible. And then the protopsychiatrist asked, basically out of nowhere, "Would you rather have HIV or hepatitis C?"

I was struck equally by the randomness of the question and the immediate and unanimous response: HIV, to a person.

Some of you may be surprised to read that, but this is one of those times when people with some medical training might see the world a bit differently than everyone else. When it was first discovered in 1983, human immunodeficiency virus infection (HIV) was a devastating disease, striking down the most vulnerable and marginalized portions of our population. The destruction wrought was terrifying and tragic. But by 2006, powerful HIV treatments were available—treatments that would change HIV from an inevitable death sentence to a chronic disease, terrible but manageable. Hepatitis C, though, was still terrifying to those of us who saw what it could do. Hepatitis C led us to witness, bar none, the most gruesome deaths we had ever seen.

Mr. Jones had contracted hepatitis C in the late 1990s through IV drug use. While 20 percent of people clear the initial hepatitis C infection on their own, Mr. Jones was one of the majority in whom the virus becomes chronic. With its tendency to infect liver cells, the virus slowly damages the organ, replacing the metabolically active cells with useless scar tissue. The scars begin as small nodules and spread slowly throughout the liver, and as they do, the functions of the liver decrease. This was the case with Mr. Jones, who had advanced cirrhosis when he was brought to the hospital delirious, combative, and vomiting blood.

An ultrasound would show that his liver had been nearly entirely replaced by scar tissue, and without the organ functioning, Mr. Jones had no way to produce the proteins that help blood clot. We would confirm this with laboratory tests, but the evidence was all over his face, clothes, and the floor of the ER. He had ruptured what is known as an esophageal varix. Over time, blood flowing from his intestines toward his heart had met too much resistance in the scarred liver, forming new pathways of drainage around his esophagus (the tube that connects the mouth to the stomach), creating big, bulky, and fragile veins around its walls. One of those veins had burst, and the lack of clotting factors meant it would keep bleeding until we found a way to stop it. The blood, draining into the stomach before Mr. Jones retched it up, was partially digested by stomach acid, turning it into a reddish-yellow soup filled with small black dots. We call this "coffee ground vomiting," which is a disturbingly accurate description and a very bad sign.

The first step to try to save Mr. Jones was to secure his airway. With this much blood coming up, he was at high risk of inhaling some of it into his lungs, compromising his ability to breathe. We sedated him, put a tube down his throat, and hooked him to a mechanical ventilator. Blood gurgled from around the tube, prompting us to quickly place another tube down his nose and into his stomach to suck the blood out. A container on the wall let us keep track of how much was removed. The speed with which it was filling with blood, "coffee grounds," and bile was unsettling.

The gastroenterologists performed an endoscopy (inserting a light and camera into Mr. Jones's esophagus) at the bedside to try to locate the ruptured esophageal vein, but they couldn't get the bleeding under control. With no other options, they placed an inflatable tube down Mr. Jones's throat and blew it up to try to squeeze off the bleeding. The tube was held in place by an inflated bubble in the stomach on one end and kept under tension by being tied around a football helmet that was placed on Mr. Jones's head. Having never seen this procedure before, the surreal nature of it all—the tubes, the blood, the football helmet—was almost impossible for me to process. It's an image that has stuck with me all this time, and one I know I'll never forget.

For the moment, this therapy worked. The bleeding slowed, and Mr. Jones was able to be transferred to the ICU. He was stabilized but still critically ill, requiring around-the-clock transfusions of red blood cells, platelets, and clotting factors. Within a day, it was clear what the end of his hospital course would be. He was too sick to survive a liver transplant operation, and his vital signs were trending in the wrong direction. Eventually, in consultation with the medical team, his family decided to stop aggressive interventions. He died shortly thereafter.

Nurses are angels for many reasons. But one that doesn't often get talked about is how they care for a body after death, removing tubes, lines, and bandages, cleaning the face and the bed, so that the family can say one last goodbye before the body is transferred from the hospital bed to the morgue. In my brief time with Mr. Jones, the healthiest he'd ever appeared was shortly after he had died, lying in white sheets without blood, his face clean, his eyes closed, and without a single piece of plastic in his body.

So, yes, I would rather have HIV than hepatitis C.

But that was 2006. At that time, if you were among the 80 percent of people who went on to have chronic hepatitis C after the initial infection, your only treatment option was a combination therapy of interferon and ribavirin. This drug cocktail was incredibly hard to tolerate. Treatment with interferon required a weekly injection, which would often be given on Fridays because the side effects were so severe that patients would not be able to work for a couple of days afterward. The duration of treatment was forty-eight weeks. If you stuck with it—and few could—the chance of cure was still only about 50 percent.

And then a breakthrough occurred. In 2013, the FDA approved a new drug—sofosbuvir—for the treatment of hepatitis C. Part of a new class called "direct-acting antivirals," sofosbuvir and related drugs changed the landscape of hepatitis C almost overnight. In fact, the desperation for an effective hepatitis C treatment was so high that sofosbuvir saw the fastest adoption of any drug in US history.

These drugs were game changers. Instead of forty-eight weeks of injection-based therapy with terrible side effects and a 50 percent success rate, the direct-acting antivirals required taking one pill a day for twelve weeks. They had minimal side effects. And the cure rate was over 95 percent. Nearly overnight, hepatitis C had been changed from a disease that kept medical students like me up at night to one that could be cured almost as easily as strep throat.

There is no doubt that these drugs were a breakthrough, but what often gets lost is the long history of failure that existed prior to their development. It was the mid-1970s when NIH researchers concluded that a new hepatitis virus (then called "non-A, non-B hepatitis") was causing liver disease among people who received blood transfusions. Attempts to isolate the virus failed for more than a decade. It was 1989 before the virus could be fully characterized.

The first treatment for hepatitis C, interferon alpha-2b, was approved in 1991 and had a paltry 6 percent cure rate. The introduction of ribavirin in 1998 upped cure rates to 50 percent, but the next fifteen years saw multiple drugs fail to move the needle on this devastating disease. In short, the history of hepatitis C is one of decades of failures, leading to a spectacular success.

Such is the case with virtually all successes in Medicine. They are built on the backs of strings of failures—failures that are rigorously analyzed for signs of what went wrong and what can be improved, which lead to more failures, and more improvements, until, eventually, there's a breakthrough. This is why the idea of an "incurable" disease is nonsense. Some diseases just haven't been cured yet.

IF YOUR MAJOR source of medical news is the mainstream media, however, you may think that, with rare exceptions, medical research is simply a parade of astounding successes: SCIENTISTS HAVE DISCOVERED THE KEY TO TREATING ALZHEIMER'S DISEASE; NEW TREATMENT MAY CURE MULTIPLE DIFFERENT TYPES OF CANCER; FOLLOW THIS

SCIENTIFICALLY PROVEN METHOD TO LOSE THOSE EXTRA POUNDS. The reason medical successes (or studies that can be spun to *sound* like they are successes) get so much press attention is because they are exciting. They generate engagement. The press is giving its readers what they want to hear.

No one would be interested in more realistic news articles that accurately characterize the state of affairs in Medicine: DRUG WITH PROMISE TO TREAT ALZHEIMER'S PROVES WAY TOO TOXIC FOR HUMAN CONSUMPTION; CANCER TREATMENT EFFECTIVE IN PETRI DISH, NO DATA ON HUMANS YET; DIET LEADS TO, AT BEST, TRANSIENT WEIGHT LOSS. This is not entirely the fault of the press—*I* don't like writing about negative medical studies either—but it does create a very false perception.

Failure isn't the exception in medical research; breakthroughs are the exception. Safe, effective treatments are the exception. By the time a medication reaches wide use, it has passed through a veritable gauntlet of potential failure points, from lack of efficacy to unacceptable toxicity. Our pharmacopoeia is a book of gold-medal winners, survivors, the cream of the crop. Which is why physicians are so skeptical of new treatments, especially those that have not gone through the rigorous vetting process of clinical trials and FDA approval.

We believe that there are cures for every disease (even if many are undiscovered), but we also know that those discoveries come at a cost—that long string of failures. Cures don't happen out of the blue.

The Cure May Exist, but Not Where You Think

If there was a true turning point in the coronavirus pandemic, it was December 14, 2020, when Sandra Lindsay became the first person in the United States to receive a coronavirus vaccine outside of a clinical trial. The success of the coronavirus vaccines was beyond what any of us dared to hope for. In fact, in October 2020, I wrote a piece for *Vox* preemptively arguing that a vaccine with 50 percent efficacy would be an incredible achievement and could turn the tide of the pandemic (if people would actually take it). That we got *multiple* vaccines with greater than 90 percent efficacy bordered on miraculous. But, as we have already discussed, vaccination rates, while initially robust, more or less stalled in the United States in the face of mounting vaccine hesitancy, outright misinformation, misjudgment of the magnitude of the risks and benefits, and the proliferation of purported "miracle cures" for COVID.

The appeal of a miracle cure for COVID-19 was obvious. The disease was terrifying: Airborne and spreading rapidly, there seemed to be little protection against infection, short of locking yourself in your own home. The outcomes of infection could be severe, including death, and, perhaps more concerningly, the variability in outcomes was striking. Older people died more often from COVID, but young people died too, as did people with no comorbidities. It was difficult for even the healthiest among us to feel safe. But if you believed that a widely available pill could cure COVID, you could feel safe again. *Sure, COVID is everywhere and killing people,* you could tell yourself, *but here is a secret weapon that will protect me and those I love.*

The first such "cure" was hydroxychloroquine. Others followed: melatonin, zinc, azithromycin, ivermectin. They all had three things in common: They were cheap, they were widely available, and they were old. These were medications with a *long* track record and a well-defined list of side effects. Our collective experience using the drugs, for decades or more, was often touted by proponents in contrast to the "new" vaccines, whose long-term effects would not be known for years to come. (As a side note, virtually all vaccine side effects occur within one to two months of administration, because the substance of the vaccine is gone from the body by that point in time. It is very unlikely that new side effects would occur one year, two years, or ten years after vaccination.)

These potential cures all had some biologic plausibility, to be sure. But, again, that is only the start of research, not the end. What these cures lacked was high-quality evidence that they were actually effective. Remember, the most common outcome of any new treatment, by far, is failure. We can't *assume* anything works without being tested. The history of medical research is littered with promising drugs that simply didn't work.

To be fair, pharmaceutical companies had no interest in conducting studies of drugs like hydroxychloroquine or ivermectin; proving that a generic medicine that costs ten cents a pill is a COVID game changer is not in the shareholders' best interest. But multiple studies *were* conducted by independent groups and government funders, including the NIH. And none found that these drugs had a significant impact on the course of COVID-19 infection. (A few studies of ivermectin showed dramatic effects but were later shown to be fraudulent—see chapter 9.)

It didn't matter. The promise of a miracle cure, one that was cheap, widely available, and virtually side-effect-free proved irresistible to many. And, of course, if a miracle cure did exist, why would you get vaccinated? Motivated reasoning defends your status quo, and *not* getting vaccinated for COVID-19 is what we've been doing all our lives. Thus, even when definitive studies were published showing no effects of the miracle pills on COVID-19, there was little dent in enthusiasm. The public wanted a breakthrough.

An entire media ecosystem grew up around claims that these drugs were effective. Pundits, including some doctors, appeared on national television and even before Congress, touting their promise. Individuals promoting the miracle cures got much more airtime than more sober-minded analysts, who pointed out that there was no evidence supporting the claims. Poorly conducted studies that hinted that the drugs might work were posted to Facebook and Twitter, and went viral before fact-checkers could point out that higher-quality studies had found no effect. And some individuals, distrustful of experts and scientists, were quick to believe that a conspiracy was afoot to deprive the world of these cures to promote a toxic vaccine.

All of this had measurable effects. For example, a study published in September 2020 found that a belief that hydroxychloroquine could cure COVID-19 was the single most powerful predictor of vaccine hesitancy—outstripping age, sex, socioeconomic status, and even political affiliation.

I was not surprised that these drugs did not have robust efficacy against COVID-19, because the chance that any given off-the-shelf drug would be a miracle cure for COVID-19 is essentially zero. But in the face of a scary disease, the *desire* for a ready-made solution is incredibly high. We want it so much we forget that failure comes first.

The same phenomenon occurred during the other great pandemic of my lifetime: HIV.

The Temporary Cure for HIV

If you were alive in 1985, you'll likely remember the fear associated with the emerging HIV pandemic. It was in 1985 when Ryan White, a boy who was infected via a blood transfusion, was banned from attending his public school in Indiana. The disease was uniformly fatal, and there was no cure. (Ryan died a few months after his eighteenth birthday.) But a paper published in the journal *Medical Hypotheses* suggested something remarkable—that high doses of vitamin C could trigger remission among individuals with HIV. Linus Pauling, winner of both the Nobel Peace Prize and the Nobel Prize in Chemistry (but not a medical doctor), was enamored of the hypothesis and spent much of the latter part of his career touting the use of vitamin C for HIV, cancer, and other maladies.

While the protocol was never widely adopted by the broader medical establishment and was never proven to be effective, many who suffered from HIV and AIDS embraced it nonetheless, in some cases forgoing effective antiviral therapies like AZT, which was approved in 1987. Choosing vitamin C, a benign chemical with no effect on HIV, over AZT, would have been a fatal choice, but it was an understandable one. It employed the same logic that leads people to forgo vaccination for COVID-19 in favor of the promise of a quick cure should they become ill.

Knowing just how common failure is in Medicine may help reset our expectations when promising (but unproven) treatments are announced. But how *do* drugs fail? There are numerous failure points along the development path. I often think of this path as an obstacle course, or a gauntlet. Only the strong survive.

The New Drug Gauntlet

New medications are run through a large series of studies before they are even allowed to be tested in humans. Most of my students are shocked when I tell them just how unlikely it is for a medicine to reach the human-testing stage. Imagine you have a substance that you believe will help people with the medical condition Alzheimer's disease. Given the lack of effective treatments, this would be a huge boon to the world if it actually slowed the progression of the (for now) irreversible disorder. You've done computer simulations suggesting the substance is safe and done tests in petri dishes on human neurons (the cells inside the brain). With some promising data, you move on to animal studies, showing benefits in mice, rats, and even some nonhuman primates like the rhesus monkey. Finally, you are ready for your first real human study: a phase 1 trial.

These early studies enroll just a handful of people and are used to ensure that the new treatment is safe and to determine what the dose should be (documenting the efficacy will come later). By the time you have hit phase 1, your drug has already shown substantial promise. The vast majority—999 out of 1,000 substances under investigation—never make it to this point. Even so, only about five out of one hundred medicines that make it to their first live human study will ever get FDA approval and find themselves on pharmacy shelves. If you had to make a bet on any new drug, the smart odds are always to bet against. And yet we keep searching, like a fantasy hero on some perilous quest, because we know something will work eventually.

I've often wondered if the cultural heritage of fantasy and fairy tales gives us a false sense of the likelihood of cures. In those narratives, there is often a substance out there that can cure the affliction affecting the main character or someone the main character cares about. Aesop tells the story of a sick lion, who, according to a fox, would be cured only by the skin of a wolf. Snow White, in the Grimm brothers' original work, is saved by the seven dwarfs on two separate occasions (prior to the infamous poisoned apple), thanks to their quick remedies for the various poisons the evil queen has concocted for her. Through the centuries, variations of the Fountain of Youth or the philosopher's stone speak to the innate human hope that there is something out there that can free us of the existential dread of our own frailty and inevitable death. We believe these cures are out there because we have been told, from the time we were old enough to listen, stories of cures and remedies, revivals and resurrections. This is a hope that truly springs eternal.

And it's not even wrong. The cures *are* out there. They are just substantially more difficult to find than we were led to believe from storybooks. Even large pharmaceutical companies, with hundreds of scientists and access to the most modern methods of screening through millions of compounds, fail much more often than they succeed. When faced with any disease, novel or not, we need to test our best guesses but be prepared for the reality: We almost never guess right the first time around.

Failure Is a Good Thing: Doctors and Pascal's Wager

Earlier in the book, I wrote about how true data can be interpreted differently by different people—particularly by doctors (who think in terms of populations) and patients (who think in terms of individuals). But the prerequisite for that discussion is that there is data to interpret. What about when there isn't data yet? For any given disease or health problem, there is a time before we know whether or not a given treatment will work: before it has been tested in the robust causal framework discussed in the last chapter. In those situations, we find ourselves in a scientific twilight zone, where there is no conclusive proof that a treatment *will* work, but no conclusive proof that it *won't* work either. A question that comes up from patients all the time in these twilight areas is "Why not try?"

The pat answer is "Side effects." Doctors are quick to trot out the risk of side effects to avoid prescribing a medicine without a strong evidence base, but, to be honest, I think we overplay this card. While most medications do have side effects, most side effects are mild and self-limited. And if they develop, you can usually stop the drug and you're no worse off than when you started (although in the US healthcare system, you are likely to be poorer).

Doctors also take an oath to "first, do no harm," which we take seriously. That part of the Hippocratic oath leads us to be a relatively conservative bunch. We do not like to take risks with our patients, and any medication without a solid causal link to improved outcomes constitutes some risk (however minimal) without obvious benefit. But I think the risk-benefit language is not entirely justifiable. After all, people aren't asking us to prescribe arsenic. In general, these issues revolve around a medication or substance with a relatively benign safety profile and a biologically plausible (but not yet proven) mechanism of benefit. Take, for example, the push to prescribe melatonin for COVID-19.

Melatonin is a hormone released by the brain that helps to regulate our sleep-wake cycle. Melatonin supplements are relatively safe (though they may make you drowsy), cheap, and sold over the counter. Early in the pandemic, some researchers suggested that melatonin could be a treatment for COVID-19, thanks to some of its anti-inflammatory properties. This is a reasonable hypothesis, with some biologic plausibility. Many patients were asking "Why not?" Why not take this cheap, benign drug? If it works, great. If it doesn't, no harm done.

This argument is a variation of "Pascal's wager." Blaise Pascal was a seventeenth-century philosopher, mathematician, and all-around genius. In his posthumously published work *Pensées*, Pascal reasoned that humans should believe in God because if God does not exist, their loss would be minimal, whereas if God does exist, they could reap infinite rewards. Conversely, if they do not believe and God does turn out to exist, they could face infinite punishment. The formulation is one of the earliest examples of decision theory, though I'm not sure God would appreciate being believed in simply to hedge one's bets.

There are many philosophical arguments against Pascal's wager, but the one relevant to Medicine was best expressed by the eighteenth-century philosopher Denis Diderot, who reportedly stated, "An imam could reason the same way." This argument from inconsistent revelations essentially boils down to the fact that if you believe in the *wrong* God, believing might be *worse* than nonbelief, particularly if the real God is the jealous type.

Pascal's wager breaks down the same way when it comes to Medicine. The argument to take melatonin for COVID-19, for instance, essentially says that if you turn out to be wrong, you lose little, but the rewards for being correct are potentially lifesaving. The argument from inconsistent revelations applies here as well. By choosing to take melatonin, you are de facto choosing *against* taking some other medication that may be lifesaving itself. And if biologic plausibility is our guide, there are hundreds—perhaps thousands—of medications to choose from.

In that context, some might be tempted to take a "kitchen sink" approach, as millionaire podcaster Joe Rogan did when he was diagnosed with COVID-19. "All kinds of meds," he told his audience when describing what he took. "Monoclonal antibodies, ivermectin, Z-Pak, prednisone—everything. And I also got an NAD drip and a vitamin drip."

You don't have to be a doctor to know that this is not a great way to practice Medicine. Taking a mishmash of disparate drugs poses substantial risks, due not only to the adverse effects associated with each individual med, but also to the increased risk of medication interactions. We also learn nothing from this approach—Joe Rogan survived his bout with COVID, but we have no idea which of these drugs (if any) made the difference.

We *need* a filter to take the large number of possible medications down to some small number that are reasonably, causally, linked to an improved outcome. Otherwise, as doctors, we'll be more or less prescribing at random or at the whim of whatever new drug has captured the zeitgeist or has the best marketing.

That filter is failure. I often imagine the potential treatments for a disease written out on some ethereal scroll: compound after compound in beautiful calligraphy, stretching off into the distance. This is the scroll of potential. Somewhere, among this list of one thousand, or ten thousand, or one hundred thousand substances—some of which do not even exist yet—there really is that magic elixir we heard about when we were five. As we study the treatments with potential, we start crossing them off the list. Those that are poisonous go first, followed by those with severe side effects that outweigh potential benefits, followed by those that are ineffective. The long arm of science scratches one name from the list, and another, and another. The filter protects us from poison, yes, but it also protects us from a tyranny of potential. Medicine by guessing isn't Medicine.

Saying no is never easy for a physician, because we really do want to help our patients. Many of us cave under pressure to prescribe unproven but benign medications, and I don't think physicians who give in to patient pressure to "just try" should be chastised, censured, or lose their licenses. This is distinct from physicians who actively *promote* therapies without evidence. Those who convince their patients to take something that has not been shown to be of benefit are, in my opinion, in violation of that most important part of the Hippocratic oath: to do no harm.

What to Do When There Is No Cure

Grand pronouncements that the arc of medical science bends toward breakthroughs may come as cold comfort to some of you reading this who are afflicted with a condition that has not (yet) been cured. Such conditions are innumerable. Some are not life-threatening but merely omnipresent and therefore life-altering. But what if you have been told that there are no more options, that you have a condition that will ultimately cause your death?

The natural tendency of those who cannot yet be cured is to

search for a miracle. We look for hope in anecdotes—googling the name of our condition and finding the occasional social media post encouragingly titled "How I Beat ____." We look to history and to alternative medicine practices, which often promise to cure what traditional medicine has deemed incurable. We trust in the comforting words of those who promise big but deliver little. I want to offer an alternative.

In chapter 3, we discussed the difference between hope and meaning, and how when the former is in short supply, the latter can make all the difference. The psychiatrist Viktor Frankl wrote, "Despair is suffering without meaning." While it is not always possible for a doctor to prevent suffering, we can work with our patients to avoid despair by helping them to find meaning.

One powerful way to create meaning is by contributing to the progress of medical science by participating in clinical trials. The easiest way to find a clinical trial is to go to ClinicalTrials.gov. It is a requirement for all clinical trials to be registered there, and its search engine allows you to sort by disease, treatment, and location, among other categories.

Those who try to recruit patients into clinical trials often imply that the drug under investigation is likely to succeed. They are not trying to hoodwink you—it's just that researchers choose the studies we do because in general we believe that the intervention or drug being tested will work. We're wrong most of the time, as this chapter has illustrated, but if we didn't believe in it, it would be hard to go to work in the morning. We understand, rationally, how likely failure is, but we also know that eventually there will be a success, and so why shouldn't it be ours?

The unfortunate truth is that the majority of people who take part in a clinical trial won't benefit directly. Most placebo-controlled studies randomize in a 1:1 ratio, so half of those enrolled won't even receive the drug under investigation. But even if you are lucky enough to get the study drug, there is a good chance that it will have no effect.

Many people participate in these trials because they hope they will get lucky—that they acquired their incurable condition at just that point in history when the condition becomes curable. And I don't disparage them that. There was a first person who was cured of hepatitis C, and acute myelogenous leukemia, and syphilis, and breast cancer, and hundreds of other maladies that would have struck our ancestors down in the prime of their life. But that the first cure will be you is not a likely situation. Rather, I encourage you to participate in clinical trials to provide meaning when hope can't be found, to contribute to that long and necessary string of failures that comes before that great success. The cure is out there. It will change the lives of our children or our children's children, and you, by volunteering yourself, will be a part of it. While you might not be cured, you will be a part of the cure.

If you choose to find some meaning in this way, I thank you on behalf of all of us who try, and fail, to advance medical science, and on behalf of the citizens of the future, who will be living in a world of cures.

CHAPTER 8

Pharma

I DIDN'T THINK THE medications known as SGLT2 inhibitors would work. The new diabetes drugs—each major pharmaceutical company seemed to have one in its pipeline—all had the same idea: Cause individuals to pee out extra sugar. The key insight that led to the development of these drugs was that the kidneys, whose job it is to filter toxins and other nasty metabolites out of the blood, *also* filter out the sugar. This is a terrible thing, since for much of human history, peeing out sugar, and the calories it represents, would be a huge waste of important resources. Thankfully, the kidney tubules have a special type of transporter that is highly selective for sugar. These transporters watch the urine floating by, full of all those nasty toxins and metabolites, and grab the sugar for return to the bloodstream. The SGLT2 inhibitors stop these transporters. The sugar gets filtered, flows freely through the kidney tubules, to the bladder, and then . . . well, out to sea.

The drugs had a clever mechanism, and my prediction was that while they might modestly lower some of the sugar-related metrics we measure in people with diabetes, they wouldn't make a major splash in terms of key outcomes like death. I was wrong.

Starting around 2016, study after study of the drugs came out showing similar conclusions: SGLT2 inhibitors didn't just improve sugar numbers; they reduced the risk of all-cause mortality by about 15 percent. And though the drugs started out as a diabetes treatment, newer studies suggested similar mortality benefits in patients with heart failure, even if they did not have diabetes. These were game-changing results. We had never seen a drug class emerge that had so strong an effect on saving lives among these patients.

And yet doctors weren't prescribing them. Or rather, they were, but slowly, and not in the numbers that you would expect of a new breakthrough medication with the real potential to save lives. This lag in new medication uptake is well described, and pharmaceutical executives absolutely detest it. Look at it from their point of view: They have spent hundreds of millions of dollars developing a new drug and putting it through the rigors of formalized testing in expensive randomized trials. They have received the stamp of approval from the FDA, and now, just when all that work is supposed to pay off, the drug doesn't get used. They pull out all the stops at this point, advertising directly to consumers, begging them, "Ask your doctor about..." and "See if this is right for you." And yet it takes years, sometimes decades, before the drug becomes what we would consider standard of care.

There are a few reasons for the lag. Doctors tend to trust what we know and are slow to bring new medications into the fold, regardless of how great the data is. And, frankly, many doctors are not familiar with the latest research on new drugs. In a day full of running from clinic room to clinic room, seeing patients, we don't always have time to flip through the *New England Journal*.

It turned out my lab had a potential solution to this problem. For the past ten years or so, I have been running experiments to figure out how we can get physicians to make good choices for their patients. Formally, we call this area "clinical decision support." The idea is that when a physician is seeing a patient, we create a tool that looks at the patient data and provides advice on best practices. We started with supersimple stuff: "Hey, this patient has worsening kidney function. Maybe you should perform a urinalysis." We worked with physicians to make these prompts functional and unobtrusive. We even studied their effects using randomization (which at this point shouldn't surprise you), to make sure that we could determine if our little prompts actually caused changes in physician behavior and, even better, patient outcomes. My lab had received several NIH grants for this work, and we were eager to expand into new areas.

But until I got a phone call from one of my cardiology colleagues on a sunny day in April 2018, I had never really thought about working with a pharmaceutical company on this issue.

"They are superinterested in this, man," my colleague said. He had followed our work peripherally for a few years, and at a recent event with a pharmaceutical company that I shall not name (for reasons that will soon become clear) he mentioned to a couple of the company's executives the work we were doing: Could we prompt physicians to prescribe an SGLT2 inhibitor using an electronic alert system? Their ears quickly perked up.

I was ambivalent about the idea. On the one hand, it was clear to me that these drugs were truly good medicine. The studies were well executed and compelling, and the drugs treated a population (those with diabetes or heart failure) without a lot of good options. When I saw patients like this, I would prescribe an SGLT2 inhibitor. And I saw the same data the pharmaceutical companies saw: Not enough doctors were using the drugs. Getting more docs to appropriately prescribe these drugs would save lives.

But the motivation of the company here gave me pause. Yes, more prescriptions would likely save lives. At the same time, more prescriptions would also generate more revenue for the pharmaceutical company. Our interests were not exactly aligned, and yet they were both somehow pointing in the same direction. I decided to take a call.

Over the next few months, I had nearly weekly calls with this pharmaceutical company as I and my team wrote a formal study protocol to evaluate whether these types of prompts and alerts could increase the prescription of SGLT2 inhibitors, and consequently reduce hospitalizations and deaths among those so treated. The company would pay the bills for the study, but, importantly, the execs let me know that the running of the study would be entirely in my hands—the data would be analyzed by my team, the manuscript would be written at Yale, and they would not interfere in any way with its publication, regardless of the results.

We built a prototype of the alert fairly quickly. At the time, there were three SGLT2 inhibitors approved for use in the United States. Our alert informed the physician that their patient may benefit from an SGLT2 inhibitor and that if they agreed, they could prescribe one of them. We created a nice display with checkboxes allowing the physician to pick whatever drug they or their patient wanted. (Of course, they could ignore the recommendation entirely as well.) This is where the trouble started.

"Oh, you have all three meds on there?" one of the execs said.

"Yeah," I said. "There are only three that are FDA-approved right now."

"Sure, right. But we thought it would just be *our* drug."

It all clicked for me. While I wanted patients to be treated with *an* SGLT2 inhibitor, this company, who was going to pay the bills for the study, wanted patients to get *its* SGLT2 inhibitor.

I tried to explain that this was a bad idea. First, I appealed to science, stating that the evidence for all of these drugs was very strong, and that there had been no direct head-to-head comparisons yet. Since we didn't know whether any one SGLT2 inhibitor was better than any other, we shouldn't preference one over another.

That didn't work.

I then tried the logistical approach: "You know, different insurance companies will preference one drug over another. We don't want patients to pay more when they could get a similar drug for cheaper."

No dice.

Finally, I tried to hit them where it might hurt the most: public image. "Look, if we only recommend this one drug, *your drug*," I said, "and right there in the alert is a disclaimer that the study is funded *by you*, don't you think that would look terrible? I mean, one newspaper story about this study, and people would be outraged."

In the end, after about six months of preparatory work, the company decided not to fund the study. My lab never saw a dime, and my first experience "working with pharma" left a bad taste in my mouth.

It also led me to a new understanding. The companies are not evil. They are not malevolent. They are merely self-interested. That they are self-interested doesn't mean they are wrong or that their products won't help people. It just means that their goal of helping people is *in service* of the goal of helping themselves and their stockholders. Pharma is a powerful force, one that does not always push in the direction we want it to. But it is nevertheless a force we can harness for good—if we are careful.

Why You (Probably) Hate Pharma

In 2019, a Gallup poll found that of twenty-five major industries in the United States, the favorability rating of the pharmaceutical industry was dead last. Pharma was rated lower than healthcare, lower than the legal field, even lower than the federal government itself (which, in the United States, is really saying something). The disdain for pharmaceutical companies runs deep, and it isn't too hard to understand why. US drug prices are the highest in the world—often by orders of magnitude, a result of the legal inability of the federal government to negotiate drug prices for those with public insurance like Medicare or Medicaid.

That anticompetitive law remains in place at least in part due to lobbying dollars. In 2021, the pharmaceutical industry spent more than $350 million lobbying Congress. The next closest industry, electronics manufacturing and equipment, spent around half of that. The oil and gas industry—which we often associate with shady backroom politicking—spent just over $119 million lobbying in 2021. If you are in a position to make or change laws, chances are you are getting campaign checks from pharma. Two-thirds of Congress members received pharma donations in 2020 alone.

What does all that lobbying get you? It gets you $1.2 trillion. That is the amount of money the pharmaceutical industry made around the world in 2021. And $500 billion of that was made by US pharmaceutical companies alone. It is hard to describe how much money that is. With $1.2 trillion, you could build 170,000 elementary schools, or send nine million kids to a four-year college, or buy 1,000 Buckingham Palaces or 923 nuclear aircraft carriers.

The industry is very big. But that, in and of itself, is not necessarily a bad thing. By revenue, the pharmaceutical industry is actually smaller than the auto industry and the oil industry. It's smaller than the finance industry. In fact, the worldwide revenues of all the US pharmaceutical companies combined are about on par with what Walmart makes in a year.

Pharma enjoys a privileged status in the United States compared to the rest of the world, and not only due to the fact that the United States is the only country that doesn't negotiate the cost of the drugs it buys. Only two countries in the world, the United States and New Zealand, allow direct-to-consumer advertising of prescription pharmaceutical products. It's a sad bit of American exceptionalism that this country is one of the only places on Earth where you might hear "Ask your doctor if Viagra is right for you" coming from your TV screen. Although you may think these advertisements are harmless, or even humorous (with their running-through-fields imagery underscored by the rapid-fire listing of side effects like diarrhea, suicidality, and death), the industry would not be running the ads if they weren't successful.

When you see these advertisements—the happy couples dancing through the fields, or holding hands in his-and-hers bathtubs, or opening the window blinds to a swift sunrise—you can't help but be frustrated, particularly if you are spending money to take some of these medications. *Surely,* you may think, *they would be cheaper if the company didn't spend money on advertising.*

When criticized for the high prices of approved drugs, the industry is quick to point out the staggering costs of research and development as a major driver, but a report from PharmacyChecker, using publicly available records, found that in 2019 the pharmaceutical companies Eli Lilly, GlaxoSmithKline, Pfizer, Novartis, and AbbVie *all* spent more money on marketing and sales than on research and development. So, in short, yes, your drugs are more expensive because you are paying for those ads.

The excessive cost of medications is not just an abstract concern, another source of inexorable crushing medical expenses that threaten the entire system. The costs lead to direct, personal patient harm. A 2019 poll found that 29 percent of adults have skipped a prescription medication because of the cost. According to the West Health Policy Center, 112,000 Medicare patients could die prematurely every year due to being unable to afford lifesaving medication—a number higher than the number of yearly deaths due to diabetes. This is a problem. You may have created a lifesaving drug, but it won't save all the lives it could if no one can afford it.

The high price of medications also eats away at the doctor-patient relationship. You may trust that I am trying to help you, but that trust can be shaken when I prescribe you a drug that costs hundreds or even thousands of dollars a month to get. Why do we do it, then? The real answer is that in many cases we simply have no idea how much a drug will cost you. A study published in *PLOS Medicine* found, for example, that physicians' estimates of drug costs were accurate less than 50 percent of the time. We tend to overestimate the cost of inexpensive drugs and substantially underestimate the cost of expensive drugs.

This isn't purely out of ignorance. It's the fact that, depending on how your insurance company has negotiated with the drug company, two patients may pay wildly different prices for the same medication. Those SGLT2 medications that save lives for people with diabetes? A patient who has already hit their deductible may get them for literally zero dollars. A patient who has yet to hit their deductible would be paying around $500 a month for them, and that's *with* a coupon card. Because the variance in cost is so wide from patient to patient, we rarely even think about it. We just say, "This is a good medication—you should take it," and leave it to you to find out what you owe when you are standing at the pharmacy counter. The feeling when the sticker shock hits is understandably one of betrayal.

Patients and physicians need to demand that insurers make pricing transparent and integrated into the clinical workflow, so patients can make informed choices or at least prepare mentally for the costs they are about to incur. Price transparency would also drive prices down, as physicians could start to choose equally effective, less expensive options.

Profit Before Patients

Pharmaceutical companies pursue profits, like every other for-profit company on Earth. Our disdain for them comes from the disconnect between *what* they do (provide therapies that are life-changing or lifesaving) and *how* they do it (by charging exorbitant prices). Some of my patients are so disgusted with this state of affairs that they distrust *any* medication that comes from a pharmaceutical corporation (which is to say, virtually all prescription medications). This may feel morally satisfying, but from a medical perspective it's throwing out the baby with the bathwater. There are good drugs out there. We must be able to separate our opinion of the drug from our opinion of the drugmaker.

While charging ridiculous prices is the main way pharmaceutical companies protect their bottom line, it isn't the only way. High-profile scandals illustrate how far some companies will go to protect the profits they are making. These scandals are rare, but they're worth mentioning, since they illustrate weaknesses in the systems that have been set up to protect consumers from harmful products. Three major pharma-related scandals are highly illustrative here, in that they share a common theme—the drugs in question worked, but severe side effects were downplayed.

In some ways the thalidomide scandal is the first proper drug scandal of the modern era, the ur-scandal, and one that has shaped our drug approval process ever since. Thalidomide was developed in the 1950s in Germany, largely by ex-Nazi scientists, and aggressively marketed to pregnant women around the world as a sedative and a treatment for morning sickness. This, despite the fact that the drug had never been tested on pregnant women at all. It was soon discovered that thalidomide causes severe birth defects when taken during the first part of pregnancy (when morning sickness is most common), leading to brain damage, blindness, limb deformities, and death.

In contrast to some later scandals, these risks were understood by many physicians and researchers during the period thalidomide was being widely prescribed—some of whom argued vociferously for the removal of thalidomide from the market. But in that era, many countries had no robust mechanism to evaluate and approve drugs for sale. Fortunately, in the United States, the FDA had this authority. The pharmaceutical company that had licensed thalidomide for sale in the United States aggressively pursued FDA approval but was thwarted, largely through the efforts of one FDA reviewer—Frances Kathleen Oldham Kelsey—who stood up against the pressure and refused to allow the drug to be sold. For that act, which likely saved thousands of children's lives, Kelsey was honored with the President's Award for Distinguished Federal Civilian Service, presented by John F. Kennedy in 1962.

Thalidomide marked a cultural shift in drug regulation, with an expanded focus on safety, and a new requirement that drugs be tested in the population in which they are going to be used. Still, rare side effects occur with all drugs. No intervention is risk-free. The question is how severe the side effects are and whether the benefits outweigh the risks. But when drug approval hinges on that equation, pharmaceutical companies have repeatedly downplayed known risks, with devastating results.

In the early 1990s, a new diet drug appeared on the market: fenfluramine/phentermine, better known as fen-phen. Fen-phen was modestly effective as a weight-loss agent; people taking the drug would lose, on average, around fifteen pounds over a few months. But there was a problem. In some individuals, the drug caused a severe lung disease known as pulmonary hypertension, as well as heart valve problems. The who-knew-what-when history around fen-phen remains frustratingly opaque, but we do know that a Wyeth scientist, in 1994, questioned why the drug label noted only two cases of the severe lung disease when the study the label was based on had forty-one cases.

This is a key point. If a new drug has a terrible side effect that occurs in, say, one in a million people, you may not pick it up in the initial studies of a few thousand people. When it is approved to be prescribed more widely, these events get documented and, if necessary, the drug can be removed from the market. This is not fraud; it is simply the math of treating large numbers of people. But when severe side effects are known in advance and withheld, as it appears was the case for fen-phen, we see the ugly side of pharma: profits over people.

The FDA finally pulled the drug from the market in 1997. Wyeth netted an estimate of $308 million from sales of the drug but will lose much more than that amount in ongoing lawsuits.

History doesn't repeat, but it rhymes, and more recently, a scandal surrounding the pain reliever Vioxx echoed the thalidomide and fen-phen controversies. Vioxx, manufactured by Merck, was approved in 1999 as one of the first in a new class of pain-relieving drugs known as "COX-2 inhibitors." They worked in much the same way as other pain relievers like Advil but had fewer gastrointestinal side effects. (The older pain relievers caused stomach irritation and sometimes ulcers.)

Vioxx and its main competitor, Celebrex, were widely adopted—largely on the strength of early clinical trials, which showed they were good at relieving pain and well tolerated. But in the case of Vioxx, those early trials held a secret. The risk of heart attack was four times higher in people randomized to take Vioxx compared to those randomized to take naproxen (an older pain reliever, marketed as Aleve in the United States).

By now, the story should sound familiar. Heart attacks were under-reported to the FDA and downplayed by the pharmaceutical company. (Merck argued that what the data really showed was a *protective* effect of naproxen, not a harmful effect of Vioxx.) More studies emerged, consistently demonstrating that the drug was associated with a higher rate of heart attacks. Nevertheless, Vioxx stayed on the market for five more years and was prescribed more than twenty million times.

Writing in the *Lancet* in 2005, David Graham and his team, of the FDA's Office of Drug Safety, estimated that over the time it was in use, Vioxx led to more than eighty-five thousand excess heart disease cases in the United States and around forty thousand excess deaths. During that time, Merck racked up more than $11 billion in sales. Litigation surrounding the drug's harms, which has now been largely wrapped up by a class-action suit, has cost the company roughly $5 billion.

In none of these cases—thalidomide, fen-phen, or Vioxx—did any individual from a pharmaceutical company go to jail. And the civil penalties faced were hardly enough to deter future wrongdoing. But I want to be clear that these cases are not the norm. The vast majority of medications that make it to market are safe and effective. But these scandals remind us of the darkest side of the pharmaceutical industry, the side that seeks profits, not cures.

I also want to point out that in each case it was doctors who blew the whistle. It was good people, fighting against immense pressure, who brought us the truth. And there are good people everywhere. For the most part, those good people bring us good medicines. So what happens when the medicine really is good? What will a company focused on profits do when it develops a drug that is a true blockbuster?

Blockbusters

Drugs that are so good at what they do that an entire company can be built around them are called "blockbusters." What can pharma do with a blockbuster drug? In the United States at least, almost anything. Take epinephrine, for example. Epinephrine is clearly, unequivocally, a lifesaving drug for people with allergies severe enough to cause anaphylaxis. You probably know at least one person with a severe allergy who carries an EpiPen with them. This device, containing a measured dose of epinephrine, can be used to quickly administer the drug should anaphylaxis occur, stopping the reaction in its tracks. An EpiPen saved my friend's life one night, when she inadvertently ate some pesto. (She has a severe pine nut allergy.) This is good medicine.

It's also cheap medicine. The cost to manufacture an EpiPen, including the medicine inside, is about $35. Though the EpiPen was developed in the late 1980s, the pharmaceutical company Mylan acquired the license from Merck in 2007. At that time, you could buy an EpiPen for less than $100. By 2013, Mylan had raised the price to $265. By 2015, amid improved market share, the company increased the price to $461—at which point this one drug accounted for 40 percent of Mylan's total profits. By 2016, the price was up above $600, when public outcry led to a congressional investigation and prices were reduced.

Why was Mylan selling EpiPens for more than $600? Mylan's CEO, testifying to Congress, said the price increases were needed to help increase "consumer engagement." That response is ridiculous on its face. Were those who needed EpiPens as a lifesaving safety net really losing interest? No, I think the purely capitalist answer is likely the correct one. The price was increased to the amount the market would bear.

Does the extortionate pricing of an EpiPen mean that EpiPens don't work? Of course not. The medication works wonderfully. But it and the thousands of medications like it that might transform your life don't work at all if you can't afford to actually take them. Patients quickly lose faith in the system when the means to end or prevent their suffering is out of reach.

Their frustration often starts with pharma, but it doesn't always end there. Sometimes, physicians end up in the line of fire. After all, we are the ones prescribing the drugs that no one can afford. It is natural for frustrated patients to start to wonder whose side we are really on. It doesn't help that pharmaceutical companies can influence physicians' behavior more easily than even the physicians themselves understand.

Buying Physicians: Cheaper Than You Think

If you believe what you read online (particularly in the comments sections), you might suspect that most physicians have a steady stream of income from pharma in the form of kickbacks, bribes, and other nefarious schemes. We are called "shills" and "hype men," "sellouts" and "shysters." It's true that pharma spends an inordinate amount ($1.8 billion a year, according to a 2017 study) paying physicians for consulting, buying them lunches, that sort of thing. But these payments are spread out among a huge number of physicians—about 450,000 in the aforementioned study. That's about half of all physicians in the United States. Full disclosure: I was not one of them.

But while half of physicians are seeing *some* benefit from pharma, the median total amount received by a given physician was $201 per year. While two hundred bucks is nothing to sneeze at, the average physician is not making a living off pharma kickbacks. Many may just be getting the occasional pizza while a pharma representative goes over their product list. To be fair, some physicians are taking in huge figures from pharma—a 2019 piece by ProPublica cataloged over seven hundred doctors who made more than a million dollars from drug and device companies, largely through lucrative consulting contracts. But for the average physician, what pharma is buying with that $200, perhaps in the form of a few nice lunches throughout the year, is goodwill.

And it works. In 2021, researchers published a paper in *Annals of Internal Medicine* synthesizing decades of research to answer the question of whether payments to physicians by a given pharmaceutical company increase the rate at which those physicians prescribe drugs made by that pharmaceutical company. I can almost *feel* you rolling your eyes at this—did we need decades of research to confirm that, yes, there is a strong, consistent relationship here?

Physicians not only prescribe more drugs made by the companies that buy them lunches and other goodies, the study found; we also prescribe more branded drugs (even when a cheaper generic is available). There was even a dose-response effect (Bradford Hill would have approved!) showing that the more a physician gets paid by a drug company, the more of that company's drug they prescribe.

This is a problem. For many conditions, there are multiple treatment options. The essence of a doctor's job is to help you navigate those treatment options. If our choice between drug A and drug B is influenced by who bought us lunch last month, it means we aren't helping *you* make the best choice. Even if the drugs are equally effective and equally safe, they may not be the same price, especially if those lunches lead us to forgo cheap generics for their identical but more expensive branded versions. And frankly, we don't need that $200 a year. We can buy our own lunch.

To combat the influence of pharmaceutical companies on physicians, in 2010 the Physician Payments Sunshine Act was passed by Congress. It requires manufacturers of drugs and medical devices to report all payments to physicians and teaching hospitals to the Centers for Medicare and Medicaid Services. The information is public. You can go to OpenPaymentsData.CMS.gov/Search to look for your personal physician, your local hospital, Sanjay Gupta, or me, to see how much we've been paid and by whom.

The hope was that bringing these payments into the light would reduce their impact. However, there is little data to suggest this has worked. In fact, a 2019 study in *JAMA Network Open* suggested that trust in physicians decreased more quickly after the Sunshine Act than before. This doesn't mean the act wasn't a good idea. Rather, it's analogous to laws requiring restaurants to put their health inspection grades on the front window—we may feel a little more uneasy knowing what goes on behind the curtain.

Did the Sunshine Act at least curb payments to physicians? Not at all. They have continued more or less unabated, although some new studies suggest that the same amount of dollars is being concentrated among a smaller group of doctors.

I can tell you that in all my years of practice, I have never heard a patient mention the Sunshine Act, or that they looked me up in the database. Few patients know how to find this data (although you now do). And moreover, there is not a ton of potential for shame. If you see that your physician has received $100 worth of lunch over the past year from a pharmaceutical company, would you even bring it up? If you did, they'd be very likely to say that, sure, the pharma rep buys lunch for the office from time to time but that it doesn't influence their prescribing at all.

There is, in fact, a dramatic disconnect between the data (which shows that payments drive prescribing behavior) and physician perception of the *personal* effect of payments on their practice. Surveys repeatedly show that physicians say they do not consider pharmaceutical payments when making treatment decisions. And yet . . . the data is the data.

Talking to your doctor about these subtle conflicts of interest is particularly difficult. Your doctor could get defensive about it, jeopardizing the therapeutic alliance. Instead, I advise you to be aware of the conflicts of interest before a visit so you'll know when to ask for more information. If, for example, you see that your doctor has received payments from the company that makes the cholesterol drug Lipitor, and your doctor suggests you start taking Lipitor, you can simply ask what alternatives there are. You can even ask what would be the most cost-effective treatment. (To be fair, this is a good practice to get into regardless of whether or not your doctor has a conflict of interest!) In this way, you are not challenging the integrity of your physician. Rather, you are merely opening the conversation to make sure you are getting information about options that are not this particular doc's default.

This works really well, except in one situation—when there are no alternatives. For pharma, monopoly is not just the goal of a board game you fight with your brother over. It's a real goal. And if you end up with a monopoly, there are plenty of ways to keep it.

How to Keep Your Drug on Patent

If you're investigating a new substance or compound that may someday find its way to pharmaceutical shelves, you'll want to patent it right away, protecting the valuable invention while you work to bring it to market. Pharmaceutical companies apply for patents for a slew of substances. And as soon as a patent is issued, the clock is ticking. US patent protections for drugs last twenty years, but most drugs take somewhere from seven to twelve years to get through the testing and regulatory hurdles to come to market. That still leaves around a decade for a company to have a potential monopoly on a drug—which many of us would argue is just plenty. Patents are no doubt a good thing in terms of encouraging innovation, but all good things must come to an end.

If you are a drug company, though, the last thing you want is for your patent to end. Once it does, generic manufacturers can synthesize your drug themselves, flooding the market with a cheaper alternative. The party, so to speak, is over. As such, pharmaceutical companies do everything they can to extend patent protections for as long as possible. The main way to do that is by filing more patents. In addition to the original drug, they can patent a new dose of the drug, or a new method of administration, or a slightly different chemical structure, or they can patent the same old drug but with a new patent that identifies a new patient population who can be treated. And each of these patents can keep those pesky generics out of the market for longer, preserving their monopoly as long as legally possible.

There may be no better example of patent extension manipulation than in the world's best-selling drug. Humira, made by AbbVie, is another really, *really* good drug—a blockbuster for a reason. A potent injectable anti-inflammatory, it is life-changing for patients with rheumatoid and psoriatic arthritis, ulcerative colitis, Crohn's disease, and many other autoimmune conditions. But it isn't cheap. Humira costs about $3,000 per syringe, or around $80,000 annually, and the price has risen steadily since it hit the market. (It's about five times more expensive now than it was when it was launched in 2002.)

Since its launch, that sky-high price has generated more than $100 billion of sales for AbbVie, representing two-thirds of the drug company's total revenue. The original patent for Humira was filed in 1994, so you may have thought that generics would be available in 2014. Not so fast. As of 2018, AbbVie had submitted 247 patent applications for Humira, of which 137 were awarded. This activity has allowed AbbVie to keep generics out of the marketplace much longer. With more than $15 billion a year in sales, retaining this exclusivity through any means necessary is a trivial business decision. Generics for Humira now exist, but it will be 2023—twenty-nine years after the original patent—before you can actually use them.

Generics Won't Save You

Knowing how patent protection gets manipulated may provide the false sense that there is an easy solution here: Limit patent protections for new drugs and allow the competition of generics to drive prices down. Unfortunately, that's not really how it ends up working. Manufacturing a new generic drug is not trivial, and in the United States there is a stringent regulatory apparatus to ensure that generics are equivalent in composition and potency to their branded alternatives.

These barriers to entry mean that generics rarely enter the market the moment a patent expires. And without several options for generics, prices don't move that much. In 2019, an FDA report documented that after the introduction of a single generic competitor, the price of a medication falls by an average of around 39 percent. This improves substantially as more generics enter the marketplace. (Drugs with ten or more generic competitors saw price drops of around 98 percent.) But according to a 2017 report, 40 percent of drugs that *can* have a generic formulation have only one generic manufacturer, and the majority have two manufacturers at most.

Why? Because generic manufacturers, like the big pharmaceutical companies, go where the money is. Because of how generic competition affects pricing, it's much better to be the first generic competitor than the second or third or tenth. In some cases, if the original drug is already quite cheap, no generic companies will bother to offer an alternative.

The FDA keeps track of drugs that are off-patent and fully ready to have a generic produced but still have only one producer. As of December 2021, these include nitroglycerin ointment, estrogen gel, talcum powder, and injectable penicillin. Though these drugs are all cheap, the lack of an alternative manufacturer means that production disruptions and delays could wipe out the supply.

Alternative Medicine and Pharma Are More Alike Than Either Want to Admit

It is easy to conclude that capitalism and Medicine don't mix, or at least that *unfettered* capitalism and Medicine don't mix. (I would argue that our federal policy of not negotiating with drug companies is expressly *anti*-capitalist.) And giving our hard-earned money to a company that we feel does not have our best interests at heart rubs all of us the wrong way, however wonderful the product they may be selling. But the problem of pharma is not just limited to pharma. I will state it simply: It is a problem that healthcare is a business. And by "healthcare," I don't just mean hospitals, doctors, and drug companies. I mean all the companies out there that sell products to improve health, from herbal supplements to acupuncture to yoga.

There is no fundamental difference in the business model of Pfizer and your local naturopath. Both are concerned with making money. Both would tell you that without charging what they charge, they would go out of business and not be able to help people. Both would tell you that their products are changing lives for the better. The only real difference is that Pfizer has to spend hundreds of millions of dollars proving their product works, and the naturopath does not (provided their product does not claim to diagnose, prevent, treat, or cure any disease).

And yet our public perception of practitioners of alternative medicine is vastly more positive than it is of the pharmaceutical industry. I get the optics; the marketing of foxglove extract, reiki, and qigong is (intentionally) peaceful and patient-centric. Alternative medicine practitioners take great pains to appeal to the "natural" or "ancient" nature of the therapies, standing in stark contrast to the sterile white coldness of the pharmaceutical laboratory. And, of course, no one is paying $80,000 a year for Saint-John's-wort.

Still, in the end, I am left feeling that at least the pharmaceutical industry has data to support its claims. At least someone is minding the store, so to speak—vetting the pills for contaminants and potency, and improving therapies over time. There's a lot I want to change about pharma, but the fact that getting a new drug to market requires robust safety and efficacy testing is not one of them.

How We Fix It

I've spent much of this chapter describing the pharmaceutical industry as a (mostly) capitalist enterprise, which it is. Pharmaceutical companies have a product to sell. And in some cases it is a very *good* product—of high quality and craftsmanship, fit for its purpose and created with the latest technology. The frustration arises due to the fact that, unlike virtually all other capitalist enterprises, in which we have a choice whether or not to buy a product, when it comes to medications, many of us do not have that choice. Or, put more bluntly, our choice is to pay the price or die.

Now, these companies do realize that extortion is not a good look and, in all fairness, many have programs that allow individuals to pay lower-than-advertised prices for medications. And for those of us with good insurance (who have hit our annual deductible), those high prices don't directly affect our pocketbook. But due to the structure of the insurance industry, we *all* pay for these high prices one way or another.

To fix this, we need to fight fire with fire. As I mentioned, there is no industry that pays more money to Congress than the pharmaceutical industry. We may not have the kind of money pharma does, but we do have one thing it doesn't have: the votes. I have never been a single-issue voter, but if you were to pick a single issue to vote on, I suggest you make it the one that has the most special-interest money thrown at it. That money tells me that the industry is scared—and while Congress members do like their drug money, they like keeping their jobs more. And drug pricing reform is *incredibly* popular. In a 2021 Kaiser Family Foundation poll, 88 percent of respondents favored allowing the federal government to negotiate what they will pay for specific drugs, including 77 percent of Republicans and 96 percent of Democrats.

The rebuttal you'll hear from pharmaceutical companies is that if the federal government can negotiate drug costs with them, they will lose money (true!) and thus have less to spend on research and development. Well, as I showed you before, the drug companies already spend more on marketing than research and development anyway. Perhaps they could trim the fat a bit by cutting out that direct-to-consumer marketing they've been doing. In fact, to make it fair, why don't we just prevent *any* company selling a prescription product from marketing directly to consumers? Like almost every other country in the world.

A Fight Worth Having, but Not Our Only Fight

It's tempting to think that the key to improving health in the world is through improving healthcare, and that the key to improving healthcare is improving access to medications that prolong life or make life better. But this is a highly medication-centric worldview. Medicines cannot cure all that ails you—that's part of what makes practicing Medicine so hard. We need to acknowledge that there is so much more to improving health than improving access to medication.

Researchers have tried to put some numbers behind this idea. In 2007, well-known medical researcher Steven Schroeder, writing in the *New England Journal*, collated data to estimate the true causes of premature death in the United States. He attributed 10 percent—just 10 percent—to failures of the healthcare system (including the high cost of prescription drugs). He attributed 15 percent of premature deaths to social circumstances (like poverty and social exclusion), 30 percent to behavioral patterns (like smoking and drug use, regardless of social circumstances), and 30 percent to genetic predisposition.

In other words, we *should* fix drug pricing and access to medications. But it isn't nearly enough. It's not enough for our society, and it's not enough for any individual either. Health is not just about taking the right pills, as much as pharma wants you to think it is.

This is a fact that patients sometimes understand more than doctors. Our training is very drug- and disease-centric. In medical school, we are required to learn the mechanisms of actions, the appropriate dosages, and the side effects of thousands of drugs and chemical compounds, and to be able to know at the drop of a hat which treatment should be chosen for which ailment. It leads to the false sense that medications are the answer, when in reality they are just part of the answer.

But writing a prescription is orders of magnitude easier than getting someone to quit smoking, or eat healthier, or establish better sleep habits, or spend more time with friends and family, or love themselves. This is why it is so important for doctors to go beyond the prescription pad and remind patients that health is a big picture, and medicine is just one color.

You can help your doctor remember this too with a simple question: "What else can I do?" That question breaks us from our medication-minded stupor and opens the door to a much more fruitful discussion about true health. It affirms the importance of proper medication for diseases that benefit from medication, but embraces the idea that there is always more out there. Remember, deep down, despite the occasional lunches, we like you way more than we like pharma.

CHAPTER 9

Too Good to Be True

Fool! You're in danger of going to hell! Shame on you! Shame!

—Email in response to a video I made pointing out that a study was possibly fraudulent

IN 2001, AN article appeared in the science section of the *New York Times*, claiming something sensational. A STUDY LINKS PRAYER AND PREGNANCY, the headline announced. What followed was a brief report about a study that, if true, would have reshaped not only science but our very conception of the laws that govern the universe.

The study described a group of around 200 women who wanted to become pregnant. To that end, they were all undergoing in vitro fertilization treatments at Cha Hospital in South Korea. Unbeknownst to them, according to the three authors of the study, the women were randomized into two groups. Those randomized into group #1 received an unusual intervention: Prayer groups in the United States prayed for their successful pregnancy. Those randomized to group #2 were on their own. In the end, the study found that 50 percent of the women in the prayer group became pregnant, while only 26 percent of those in the usual care group became pregnant. The p-value associated with this finding was a tiny 0.0013. Results as dramatic as this would occur only one in one thousand times if prayer truly had no effect. These results both demanded and defied explanation.

The findings were published in the respected (and peer-reviewed) *Journal of Reproductive Medicine*. The senior author of the study, Rogerio Lobo, was the chair of Obstetrics and Gynecology at Columbia University's College of Physicians and Surgeons. All the trappings of high-quality science were there.

There were no red flags of the sort peer reviewers are trained to catch. The study was a randomized trial, the gold standard for assessing the causal benefit of a treatment on an outcome. The participants were blinded to the intervention. (They didn't know someone was praying for them.) And the effects were both statistically and clinically significant. Any intervention that doubles the success rate of in vitro fertilization would be an instant blockbuster. That this *particular* blockbuster could be delivered remotely via earnest church groups at (presumably) zero cost would be earth-shattering. As you may expect, the *New York Times* was not the only news outlet to run with this story. Stories detailing the results appeared on ABC News and in the British newspaper the *Telegraph*, among others.

Let me put this in even greater perspective. As of 2017, the success rate for IVF in the United States ranged from 13 to 43 percent, based on the age of the mother. Despite a decade and a half of scientific advances since the so-called Columbia Miracle Study, we still can't achieve the 50 percent success rate reported. In other words, according to this study, prayer wasn't just effective—it was the *single most effective fertility treatment ever developed*.

I am not a religious person. But in this case, even that does not matter. Remember, it was total strangers who were said to be praying for these women, and the women had no idea someone was praying for them. If this study were true, think of the money we could save—not just on fertility treatments, but other parts of healthcare too. I mean, sure, maybe God has a soft spot for mothers, but if prayer works for pregnancy, it should work for, I don't know... childhood cancer? Schizophrenia? Baldness?

But here's the thing: Prayer didn't work. I'm not even sure the study happened. To the best of my knowledge, no women who may have been part of this study have come forward, nor have any prayer groups. Rogerio Lobo later distanced himself from the study, saying he'd joined only after it was completed and he'd had minimal editorial input.

The story of what really happened centers around the second author, Daniel Wirth. Not a medical doctor, Wirth had a law degree and a degree in parapsychology from John F. Kennedy University in Pleasant Hill, California. A review of his prior publications reveals a litany of studies (many not yet retracted) describing the positive effects of therapeutic touch, spiritual healing, and qigong therapy. His coauthor on most of these studies was Joseph Horvath, whom he apparently met at John F. Kennedy University.

After the Columbia Miracle Study was published, greater scrutiny on Wirth and Horvath brought their house of cards tumbling down. It emerged that the two, sharing the amazing alias John Wayne Truelove, were involved in a series of scams and cons, including identity theft, check fraud, and mail fraud. In 2002, they were found guilty for a scam in which they were accused of stealing $2.1 million from Adelphia Communications Corporation, a cable company. Wirth was also found, according to the *Guardian*, to have been fraudulently collecting social security payments under the name "Rudy Wirth," possibly his father, who had died in 1998.

Why did Wirth publish this study, which, it seems, may never have even happened? It remains unclear. There was no obvious financial motive. We could speculate that he wanted to become famous or that he wanted to convince skeptics about the power of prayer. Whatever his intentions, though, studies like this harm people. After all, why would you discuss treatment options for infertility with your doctor when this study showed that a little faith would do more than any hormone treatment ever could?

I cite this example in my classes because the paper itself had all the bells and whistles we associate with good science. Had the authors picked an intervention more biologically plausible than prayer and reduced the effect size to something less incredible, it would likely still be in the literature today. How can we know if a paper we are reading is actually true?

PUT TEN DOCTORS in a room to discuss a new medical study, and you'll get ten opinions on the results. Some will be impressed, some apathetic. There will be vigorous debate about the study itself: Did the investigators use the proper dose of the drug? Did they test it in the right patient population? Do the results change what we already know? Should they change the standard of care? One thing that is almost never discussed: Is the study fake? Is the data reported accurate?

Scientific discussion of results published by Theranos, the ill-fated blood testing company, focused on the challenges of developing their blood testing technology, not whether the results were fabricated from whole cloth. Later investigations would lead to charges of fraud against CEO Elizabeth Holmes for lying about study results. Why are doctors and researchers so afraid to acknowledge fraud? Why are we willing to criticize the interpretation of study results but not the very reality of the results themselves? The simple reason is that the entire medical research enterprise is, fundamentally, built on trust. When that trust is broken by fraud, we fear that the public may become so disillusioned that they turn away from medical science entirely.

I have seen people, in clinic and online, who clearly disregard *any* medical research. All of it, they think, is some sort of propaganda—a sham in the service of higher powers like Big Pharma, or the government, or other shadowy cabals. And I have seen people who fervently believe in research studies that turn out to be lies. In short, we should trust studies that are true and doubt studies that are false. But in a world full of misinformation, it can be hard to tell the difference. Fraud pollutes the waters of scientific inquiry, harming all of us.

Caveat Emptor

So far, I've focused on the conscious and unconscious biases that color how patients and physicians interpret medical facts and make decisions about them. That's all well and good, but what if the facts themselves are wrong? To restore trust in Medicine and help people make good medical decisions, we have to acknowledge an uncomfortable truth: Some researchers lie. Some researchers bend or withhold the truth. And those actions have wide-ranging ramifications.

I should note that "fraud" is something of a loaded term and rational people can disagree on the line between fraud and "research misconduct." Completely fabricating a study is obviously fraud, but what about reporting results accurately while withholding some vital information? Or reporting on the benefits of a drug but not on the harms? Academic standards define all of these activities as frauds. In the National Academy of Sciences work *Responsible Science*, fraud is defined as "a deliberate effort to deceive and includes plagiarism, fabrication of data, misrepresentation of historical sources, tampering with evidence, selective suppression of unwanted or unacceptable results, and theft of ideas." In other words, scientists can commit fraud that is not criminal. They can commit fraud without presenting false data. Fraud merely requires the intent to deceive. I appreciate that definition because it recognizes how trust is most easily destroyed—through lies. But we must also acknowledge that proving intent to deceive can be a challenge. I do my best here to present the evidence of some questionable papers and allow you to decide on the intent behind their publication.

In my course on understanding medical research, I liken the peer-review process to buying a car based on the advertisements alone. We all know there are some truth-in-advertising laws. We trust that if an ad says the car was EPA-rated to get thirty-five miles per gallon, then probably there is some data there to support those stats. But in the end, we all want to take a test drive. We trust but verify. With medical research, the process stops at trust. We trust medical researchers to honestly report their results. Researchers rarely provide their data sets so independent reviewers can replicate their analyses. We take them at their word that they did the analyses properly. Verification is beyond the scope of peer review.

We hold medical researchers to a higher ethical standard than car dealers, and rightly so. What medical researchers are selling is new knowledge—a priceless commodity. And their product, if well-made, will genuinely benefit humankind. Conversely, shoddy science has the potential to do more damage than the rustiest jalopy.

Yet researchers are humans and subject to pressures external to the pursuit of the greater good. Though we receive no commission for our research papers, we receive grant funding to conduct that research (and the promise of future grant funding if our work is successful). We may have patents that would be particularly valuable should a certain experiment confirm a certain hypothesis. Or we may simply crave the respect of our peers, which comes from high-profile publications and academic accolades.

It is no surprise, then, that researchers often try to "sell" their rather middling research studies as something more than what they are. The previous chapters of this book have taught you to identify the signs that a study is being overhyped or overspun—"sold" rather than "told." But some researchers go beyond spin, innuendo, and speculation. There is a fundamental difference between drawing incorrect conclusions from real data and presenting data that is not accurate, because the former at least allows for an informed reader to come to a different conclusion. We need to spot the signs that the data itself may be wrong.

Is Medical Research Fraud Common?

The truth is that we have no idea how common research fraud is. Few fraudsters are forthcoming about their activities. Despite some highly publicized instances of research fraud, it's quite likely the true rate is underestimated. Governmental agencies have conducted direct data audits of clinical trials and have found that roughly two to four out of every one thousand studies have evidence of fraud. This may not seem so bad, but survey studies suggest fraud may be much more rampant. A survey of a diverse group of research coordinators reported that 18 percent had firsthand knowledge of research misconduct (which includes outright fraud, plagiarism, and other offenses) that occurred in the course of a year. At the high end of the estimates, a survey of readers of *New Scientist* magazine showed that more than 90 percent knew of or suspected scientific misconduct by colleagues. Regardless of its prevalence, fraud is doubly devastating, in that prior to discovery it may lead a field in a completely inappropriate direction, and after discovery it erodes the public trust in the scientific endeavor.

The Hidden Framework

Instances of data spinning, motivated reasoning, and cherry-picking all reside within a semilegitimate framework. It is because of that framework, which is not codified by anybody but exists in the minds of researchers and peer reviewers, that we can examine the details of a medical study as presented in prose in a medical journal and deduce whether data is being misinterpreted by the authors. This hidden framework of research has two simple tenets:

- Tenet 1: *Results* are sacrosanct.

The results section of a medical research paper is often eye-wateringly sterile. Devoid of speculation, nuance, or implication, this is the part of the paper where data is presented for the reader in as unvarnished a manner as possible.

I have been tempted to inject a bit of energy into my results sections and have routinely been shot down by peer reviewers. In one paper, I wrote:

> Concerningly, the observed mortality was substantially higher than predicted by the machine learning algorithm.

The reviewer's response was a terse:

> It is not for you to tell me to be concerned by your results.

In comparison to the staid results section, the *discussion* section of a paper is a Mardi Gras of color commentary. This is where authors speculate about their results, interpret them, and offer more hypotheses to be tested. An astute reader could skip the discussion section and be no worse off. What appears in the results section, then, is assumed to be true. For example, if the paper says that the average age of trial participants was 55.5 years, then the average age of trial participants was 55.5 years.

- Tenet 2: The *methods* are the methods.

The methods section of a study describes exactly what was done. Not only what was measured but *how* it was measured. It also describes which statistical tests were used. The promise of the methods section is that if we had access to the raw data and performed the same tests, we'd get the same results. This allows us to make statements like "They used the wrong statistical test" or "This is an unusual statistical test to choose. Why was it chosen?"

Why would a researcher who has decided to spin the results of his or her paper continue to abide by these tenets? Why would they report the results accurately, only to spin the interpretation? Fear of repercussion is one part of it, certainly. Penalties for research misconduct can be relatively mild (formal censure from your university) but in some cases are truly life-altering. In perhaps the most striking example of a scientific fraud–related penalty, Penn State researcher Craig Grimes violated the tenets when he plagiarized papers, falsified documents, and, importantly, didn't perform the research he was supposed to perform. He was sentenced to forty-one months in prison for defrauding the federal government.

But I don't think fear of Leavenworth is what keeps most researchers in line with the tenets. Researchers are humans. They want to think well of themselves, and there are certain lines that simply feel too wrong to cross. Fabrication of data is one of those lines. If I choose to use a statistical test that makes my results look slightly better than a more appropriate statistical test, I can always claim ignorance or accident. If I falsify data, there is no recourse.

It is for this very reason that identifying fraudulent studies is so much harder than identifying misleading studies. If a misleading study is a poker player trying to bluff their way out of a bad hand, a fraudulent study is a poker player using a completely different set of cards. Nevertheless, if we remain on guard against the possibility of fraud, we can drastically reduce the chances that we are taken in.

Of course, we've been taken in before, and in some cases, we continue to be taken in by frauds that have long since been revealed.

The Twelve-Patient Study that Linked the MMR Vaccine to Autism

While the Columbia Miracle Study is an extreme example of potential research fraud, the repercussions were minimal and limited mostly to embarrassment on the part of Columbia University and Professor Lobo. Well, that and the federal indictment of Daniel Wirth. But not all instances of data misrepresentation are so easily discovered, or so easily forgotten.

In 1998, a study was published in the *Lancet* that would spark a panic, start an industry, ruin a researcher's career, and give rise to endless conspiracy theories. The study was titled "Ileal-Lymphoid-Nodular Hyperplasia, Non-Specific Colitis, and Pervasive Developmental Disorder in Children," but the press would refer to it as the MMR-Autism Study or the Vaccine-Autism Study. The lead author was a British doctor named Andrew Wakefield.

Just twelve children were included in this infamous article that led millions of parents to avoid vaccinating their children and, consequently, to a resurgence of measles cases around the world. According to the paper, all twelve were happy, healthy kids until they received the measles, mumps, and rubella (MMR) vaccine. Within one to two weeks, each of them reportedly developed diarrhea and abdominal pain and began to "regress," losing the developmental milestones they had achieved. Children who could talk stopped talking. Children who could walk stopped walking. At the time of Wakefield's examination, all were severely developmentally delayed. This is an important point—it is often misstated that the kids were being studied *before* they developed their disease. The paper makes it clear that they entered the study only after the severe developmental delay had manifested.

The paper had hit all the marks of excellent science. It was published in one of the most prestigious medical research journals in the world. The authors had reported evaluating the children extensively, with tests that ranged from microscopic examination of intestinal tissue to magnetic resonance imaging. In addition, the data was compelling but not *too* compelling. While each child was reported to have abnormalities of their intestines, the authors note that the abnormalities did not correlate strongly with the degree of behavioral disturbance. The authors frankly described the limitations of the study as well, stating that perhaps the constellation of findings was due to the self-selected nature of the patients. In other words, parents of children with particular symptoms knew that Wakefield was interested in links between intestinal disease and autism, and so the patients he happened to enroll were more likely to have that symptom combination.

This was also a paper that had a compelling narrative arc; the story was truly captivating. Innocent children, on their way to leading happy, productive lives, were cut down by a vaccine meant to help them. The paper even suggested a biologically plausible pathway for the effect (missing conspicuously from the Columbia Miracle Study): "Viral encephalitis can give rise to autistic disorders, particularly when it occurs early in life." With that line, the researchers were implying that the MMR vaccine could, somehow, cause an immune reaction similar to measles infection that would manifest as both gut and brain diseases. But the researchers also added the important caveat: "We did not prove an association between measles, mumps, and rubella vaccine and the syndrome described."

On its surface, it is not surprising that this paper made it through the gauntlet of peer review. The arguments made were rather well reasoned. The subject was clearly important. The implications were compelling. In fact, the tone of understatement might have allowed this paper to fly beneath the radar, were it not for a press conference (held *prior* to publication—which is atypical) in which Wakefield, undercutting the mild language in the paper, called for a temporary ban on MMR vaccination in the United Kingdom. But it seemed there was more to this story.

In a series of articles in the *Sunday Times* from 2004 to 2010, and in the *British Medical Journal* in 2011, investigative journalist Brian Deer reported that Andrew Wakefield had been in the process of creating a business venture that would profit tremendously if a vaccine scare emerged. According to Deer, the business would specialize in diagnostic testing—looking for evidence of harm from the MMR vaccine—and the initial market would be "litigation"; the business would provide evidence supporting lawsuits against vaccine manufacturers. According to a private prospectus leaked to Deer, Wakefield reportedly stated that the company could make 28 million British pounds a year from selling diagnostic kits.

This potential revenue stream was almost certainly a substantial motivating force but was not disclosed in the *Lancet* paper. Deer subsequently identified the twelve children appearing in Wakefield's initial study, comparing the data presented in the *Lancet* to that in the actual medical record. According to his report, only one of the twelve patients had regressive autism. Only three of the twelve had nonspecific colitis. And, most damningly, not a single patient had medical record evidence confirming emergence of symptoms in the days following MMR vaccination.

Wakefield's study was retracted by the *Lancet* in 2010, which noted that "claims that [the] investigations were 'approved' by the local ethics committee have been proven to be false." But that was more than ten years after the article was originally published, and of course, by then, the damage had been done. It took more than two decades for immunization rates in the United Kingdom to recover. Measles, which was declared eradicated from the United States in the year 2000, has returned. And the narrative that Wakefield and his followers refined—fearmongering linking vaccines to devastating downstream outcomes—has become a hallmark of the resistance to vaccination against COVID-19, prolonging the deadly pandemic. To this day, Wakefield denies that there was a profit motive or fraud in the conduct of the *Lancet* study.

Seeing the Red Flags

Earlier chapters of this book explore how the careful application of logic and a few general rules of thumb can reveal whether the results of a study are worth trusting and acting upon. In the case of outright fraud, though, these techniques often fail. They fail for the simple reason that if I am going to fabricate data, I can make it fit whatever narrative I am interested in creating. I don't have to go through the rigmarole of spinning real data or drawing conclusions not supported by the data I report. I simply cook the books to get the result I want.

Logic does not always reveal fraud. For that reason, the much-vaunted peer-review process, which readily identifies incorrect research protocols, may not as readily identify outright fabrication. Peer reviewers are trained under the two tenets. When they read a manuscript, they begin with the assumption that the data being presented was actually measured. With that assumption in mind, they have already lost.

How could peer reviewers have discovered the potential fraud in the Columbia Miracle Study or the Wakefield vaccine study? Being aware of the possibility of fraud will ensure that you are not taken in by highly compelling (often widely shared and discussed) studies that will lead you into treatments that are ineffective or harmful. The techniques are simple. And they will work for you as well.

Active Skepticism

I define "active skepticism" as distinct from "passive skepticism." The latter is a prerequisite for all scientific endeavors—a state of mind that is willing to accept that the status quo may not be the whole truth. We should all endeavor to be passive skeptics. Active skepticism is harder to achieve. It is a process of deliberate consideration of alternative hypotheses to the ones being presented. In the case of a fraud, being an active skeptic means asking yourself: "Could this study be fake?"

It is understandable that few readers of the medical literature ask themselves this question. Very few medical studies *are* fake. Asking whether the study you are reading is a fraud is akin to asking yourself "Is there a ninja behind my front door?" every time you go into your house. Sure, you may be more prepared on the day that ninja is lurking, but all in all, this may not be a good use of your mental energy.

Nevertheless, I implore my students to ask themselves that question, because it induces a cognitive shift and a heightened awareness, which helps them to pull out inconsistencies that may point to the real truth.

Active skepticism is easier when the results of a study are less believable. Depending on your religious orientation, you may have been more likely to disbelieve the Columbia Miracle Study on the face of it than the Wakefield vaccine study. An individual trying to commit research fraud will not be successful if they claim to have discovered cold fusion, a cure for all cancers, or a surefire way to get all the seeds out of a pomegranate. Successful papers relying on cooked data very much resemble the Wakefield paper—the implications are often stunning (and profitable) but not so clear-cut as to draw unwarranted attention.

Know Your Authors

Part of reviewing any paper includes reading the author list that appears right under the title. We do not do this to confirm that the paper was written by the most esteemed Professor X at the prestigious University of Y. Rather, we do this to ensure that the author is writing about a field in which they are an expert and that they have no relevant conflicts of interest.

The Columbia Miracle Study had three authors. Rogerio Lobo, then chief of OB-GYN at Columbia, had impeccable credentials. First author Kwang Cha had several prior publications in the area of in vitro fertilization, though the fact that he owned the hospital where the research was conducted suggests at least some conflict of interest. But the red flags rise high with the middle author, Daniel Wirth.

Daniel Wirth was identified as a lawyer with the law firm Wirth and Wirth. Why was a lawyer authoring a paper about prayer and in vitro fertilization? Had you searched Wirth's published literature in 2001, you would have seen multiple studies in highly suspect journals, such as the *Subtle Energies & Energy Medicine Journal Archives*. This is a major red flag. Nowadays, data on prior publications is rapidly available from Google Scholar.

There were thirteen authors on the Wakefield vaccine study. It is beyond the scope of this book or your patience to review all of the authors' histories, but it is notable that Wakefield himself had no training in neurodevelopmental disorders of any kind. He was a gastroenterologist, a stomach doctor, with previous research focusing on bowel disease in children. Lack of expertise is not enough to condemn a paper—scientists are allowed to follow where the science leads—but it should raise some eyebrows and prompt further review.

Making Up Data Is Hard, Changing Data Is Easy

Pick a number between 1 and 20.

If you picked 7 or 17, you are like a plurality (roughly 30 percent) of humans who choose one of these two numbers when given this task. Humans are spectacularly bad random number generators, but if I were to make up the results of a study from scratch, I'd need an awful lot of numbers.

Let's take the Columbia Miracle Study. There were, reportedly, 219 women in that study. To report summary statistics (basic information about the patients, like their age), I have three choices.

First, I could just make up the summary statistics. The paper reports that the average age of the women in the study was 34.8 years. That number could have been pulled out of thin air. Alternatively, I could make up a random list of 199 ages and average those. Or, better yet, I could use the actual ages of actual women who underwent IVF.

The problem with strategies 1 and 2, which rely on making up numbers, is that, as I pointed out, humans are terrible at this. And computers can readily tell when a string of digits was generated by a human as opposed to a natural process. Few medical journals employ computers in this task, but they should, as the algorithms are easily implemented.

A simple example is to take the digits that appear in the tables of a manuscript and examine their distributions. A funny quirk of statistical math known as Benford's law suggests that if we were to look at only the leading digit in a data table, the number 1 will be seen more than any other number, and in a very specific proportion. If the data does not match that proportion, editors can be alerted that fraud is a possibility.

You'd think that knowledge that these techniques are out there (even if they aren't used commonly) would keep fraudsters from making up data off the top of their heads, but you should never underestimate the chutzpah of some would-be scientists. In 2005, the *Lancet* published a paper led by Norwegian researcher Jon Sudbø that claimed to link drugs like ibuprofen with oral cancer. Review of the data revealed extensive manipulation. All 908 subjects in the study were fictitious; 250 of them had the same birthday. The study was retracted.

A better way to fake results is to conduct a real study and change as little data as possible to prove your point. In the case of a randomized trial, there are two ways to do this. You can either change the exposure data or the outcome data.

Let's take the Columbia Miracle Study as an example, assuming that Wirth had access to real data on women undergoing IVF at Cha Hospital. Further assume that their data was accurately collected—the average age of these women really *was* 34.8 years, as reported. Wirth, with data in hand, had two options.

First, he could assign the randomization after the fact. In other words, he could stack the deck so that more of the women who he already knew would become pregnant were in the intercessory prayer group.

How would we determine this from the data? In most studies, it would be immediately clear from the first table of the paper, which traditionally describes the baseline characteristics of the two study arms. If the two groups are split randomly, the measurements appearing in this table should be similar across the two groups. The average age should be pretty much the same, as should a slew of other variables. This is the secret sauce of randomization, as we discussed in chapter 5.

But if assignment to each arm was not random, table 1 would be skewed. If Wirth assigned the study arms *after* he knew the pregnancy outcomes, we'd find that women who were more likely to have successful IVF (for example, younger women) would be overrepresented in the prayer arm. In the Wirth paper, this first table contains exactly two baseline measurements. This is a huge red flag. Typical randomized trials would provide dozens of baseline measures: age, race, other illnesses, medication usage, laboratory values, etc. The absence of these measures is highly suspicious.

The second way to fake these results is to change the recorded outcomes. In other words, simply find some women in the "prayer" group who didn't become pregnant and change their outcome to pregnant. It is as simple as changing a 0 in a spreadsheet to a 1. (Talk about immaculate conception.) This strategy is particularly effective because it mirrors what actually happens when an intervention works. People who would have had an undesired outcome (lack of pregnancy, death, recurrence of depression, etc.) now have a desired outcome. This type of fraud can be definitely proven only by computer algorithms, and only if all the individual data is available. Short of that, proving this type of fraud is nearly impossible, unless it is revealed by someone on the inside.

Fraud Hates a Crowd

One researcher, working alone at his desk late into the night, on his own data set, may succumb to his worst impulses and change a few outcomes here or there. But, like all conspiracies, medical research fraud becomes exponentially more difficult with more people involved. The studies most likely to turn out to be frauds are single-research-group, single-center studies. Believing in the results of this type of study is a classic high-risk, high-reward gamble. You may be on the cutting edge of new scientific understanding, but you may also be on the receiving end of an elaborate lie.

Conversely, we may put more faith in larger studies involving more researchers. Large-scale clinical trials, occurring across multiple centers, are hard to fake. They may be biased. They may measure the wrong outcome or fail to measure (or report) an important side effect. But they almost never contain made-up data—for the simple reason that such wide-scale manipulation is essentially impossible to coordinate. In several high-profile cases, individual study centers in a multicenter study have been shown to have doctored data. But there has never been a case where an entire multicenter study was proven to be fake. The Columbia Miracle Study and the Wakefield vaccine study were both single-center affairs.

Putting It Together: Detecting Ivermectin Fraud

The quote at the beginning of this chapter was in response to a video I made calling into question a study that suggested that ivermectin—an antiparasitic drug usually used to treat river blindness and scabies—had an amazing ability to nearly eliminate deaths among people with COVID. There have been dozens of studies looking at the effect of ivermectin on COVID-19, but far fewer randomized trials. Of the dozen or so randomized trials of ivermectin, the vast majority showed no effect of the drug.

To be fair, these were generally small studies, and we discussed in the last chapter that the absence of evidence of effect doesn't confirm with certainty that no clinically meaningful effect exists. But one study stood out for the profound effect size: a study led by a lung doctor named Ahmed Elgazzar at Benha University in Egypt.

According to the manuscript, which was not formally peer-reviewed, four hundred patients with COVID-19 were randomized to receive either hydroxychloroquine or ivermectin. The results were dramatic. Among those most seriously ill, there were two deaths in the ivermectin group, compared to twenty in the hydroxychloroquine group. The p-value associated with that finding was less than 0.0001. Results this weird would be *very* unlikely if ivermectin and hydroxychloroquine were equivalent. Taken at face value, then, we would conclude that ivermectin is highly effective in the treatment of COVID-19 or, alternatively, that hydroxychloroquine is a profoundly *bad* choice for the treatment of COVID-19.

But reading through the report raised red flags for me. First of all, huge effects like these are quite unusual in the medical literature. It struck me as odd that we humans would be so lucky that what amounts to almost a cure for this deadly pandemic would have been sitting on our shelves for the past forty years. Moreover, there was a biologic plausibility problem. Why would a drug that treats parasites (multicellular organisms) be effective in treating a virus that is little more than a clump of RNA surrounded by some fat? The authors had cited a test tube study showing that ivermectin inhibited COVID-19 growth in cells, but only at concentrations that were roughly one hundred times higher than what is achieved from standard dosing in humans.

Looking closer at the reported results led me to ask whether the study was actually randomized at all. Remember that if a study is truly randomized, baseline characteristics should be similar in the two treatment groups. But table 1 in the Elgazzar paper had some red flags. For example, the reported "D-dimer" (a measure of clots in the blood) of the patients in the hydroxychloroquine group was 10.2, compared to 9.6 in the ivermectin group. Using statistical software, I was able to show that a difference that large would occur only roughly 5 percent of the time if the study was truly randomized. The lymphocyte count (a measure of infection burden) was a fair amount higher in the hydroxychloroquine group as well—a difference that would occur only four out of one thousand times if this were truly a randomized trial. The ferritin level (a measure of iron stores and inflammation) was substantially lower in the hydroxychloroquine group—to a degree that would occur less than one out of ten thousand times if the study was truly a randomized trial.

My conclusion? This was not a randomized trial.

I had seen the phenomenon before, also in a preprint study purported to be a randomized trial of a treatment for COVID-19. In that case, the investigators were evaluating vitamin D, another cheap, widely available medication. The results were similarly dramatic. Of 551 patients with COVID-19 given vitamin D, 6.5 percent died. Among 379 controls, 15 percent died. A skew that weird would happen less than one in one thousand times, assuming vitamin D had no effect. The amazing results were widely circulated on social media.

I took a look at what had been posted and immediately became suspicious. The first clue was that the number of controls (379) and the number of intervention patients (551) were quite different. It's a bit unusual for a randomized trial to have vastly different numbers of patients in the treatment and control groups. Usually, coin flips are relatively balanced—if the trial has one thousand people, roughly five hundred end up in each group. Sure, some trials deliberately randomize in a 2:1 ratio or something, but the authors didn't report that at all.

Digging into the paper, I found out why the numbers looked the way they did. Patients weren't randomized to vitamin D versus control. *Hospital wards* were. There were eight hospital wards involved in the study, five of which were randomized to administer vitamin D (according to the authors), three to care without vitamin D. This might be an okay design if patients were assigned randomly to wards. But they weren't. I knew they weren't because the baseline factors for the patients in the treatment and control group were *different*. In fact, patients who ended up in the control wards had lower vitamin D levels at baseline than those in the vitamin D wards! While there may have been a minor effort toward some level of randomization, this was not truly a randomized trial.

The authors would (sort of) confirm this in a later post on the site PubPeer, in which one of them wrote, "We never say in the article that it is a randomized control trial (RCT) but we consider an open randomized trial, and an observational study." I do not really know what is meant by this.

What happened here? We can only speculate. But I tend to think that it was true that some wards in their hospital gave a lot of vitamin D, and some didn't—wards work that way. Probably, the wards were not formally randomized to this behavior. But the authors, seeing that outcomes were better on the wards that happened to give more vitamin D, may have convinced themselves that this was *close enough* to randomization. Perhaps they then wrote up their paper as if it were a randomized trial, knowing it would hold more clout that way? Maybe they convinced themselves that what they were writing was true, more or less? It could be motivated reasoning all over again. But we may never know the full story.

The vitamin D and COVID study clued me in to a new type of fraud, or near fraud, which is presenting observational data as randomized trial data. And the dead giveaway for that behavior is a wonky table 1.

I took a lot of heat for my criticism of the Elgazzar ivermectin paper—in public and in my private emails. Threats of burning in hell were not the only ones I received. So I'll sheepishly admit that I was somewhat gratified when a medical student in England named Jack Lawrence got ahold of the raw data and shared it with some other researchers. What they found was "a fraud so apparent that it might as well have come with a flashing neon sign," including the fact that "at least 79 of the patient records [were] obvious clones of other records."

I'm left again with the question of why. I understand why a huge pharmaceutical company might hide evidence that its blockbuster drug increased the risk of heart attacks and strokes. Sure, it's contemptible, but the motive (profit) is clear. But why ivermectin and vitamin D? I'm left thinking of the old legal phrase "Cui bono?" Who benefits?

Maybe it's for the fame. Maybe these docs believe so fervently that the drug works that they are willing to fabricate data to justify their beliefs—an "ends justify the means" thing. Maybe they are fooling themselves. Maybe people just don't like being proven wrong.

The Cure for Fraud

If the scientific establishment wants to take fraud seriously, there are two things it could do. Surprisingly, both of these ideas have been met with harsh criticism.

The first and most critical concept is known as "open data." Open data envisions a world where collected data sets, particularly those paid for with public funds, are openly shared. Many nonresearchers I speak with are shocked to hear that this is not the current state of affairs. When I submit a manuscript for peer review, I do not submit my research data set.

"Wait," you may say. "How do the reviewers know that what you're reporting is accurate?"

The answer, as you've learned in this chapter, is that they don't.

The open data initiative calls for journal editors to mandate sharing of deidentified study data sets with the public, to facilitate transparency. Peer reviewers could examine the data prior to publication, but, importantly, others could analyze the data after publication. Researchers, knowing that their data would be heavily scrutinized, would be much less likely to lie about the results.

The reasons some oppose open data are largely self-interested. It is very hard work to run a study, collect, and collate a large amount of data. It costs a tremendous amount of time and money. After a researcher has invested that much into a data set, how can they be expected to simply give it away for free? Researchers worry that other scientists will publish papers based on the released data, beating them to the punch and stealing the spotlight (and all-important grant funding).

These concerns, while valid, are easily allayed. A reasonable system would require a researcher who downloads a data set to agree to a set of terms. The data set could not be used for publication, for example. Or the data could be used only to verify already-published results.

More ambitiously, I and others have proposed that open sharing of data should be a criterion for academic promotion. When my promotion committee looks at my résumé, a major factor that they consider is how many peer-reviewed publications I have authored and how high-profile the journals are in which I've published. What if I could also list that my data set, collected over several years and shared openly, gave rise to hundreds of papers from groups around the world? If that metric led to promotion, we'd see a lot more sharing of data sets. We might also see novel discoveries that had not occurred to the original research team. A win-win.

The second fix for fraud is replication. Few scientists are interested in replicating studies, and funding agencies are not priming the pump either. The National Institutes of Health scores grant applications on three criteria: significance (how important the issue is), innovation (how new the research is), and approach (how valid the scientific rationale is). The "innovation" criterion is a death knell for replication studies. Replication studies are, by definition, not innovative at all. And yet they are absolutely necessary. Ending the innovation obsession at the NIH and other funding agencies could go a long way to improving the overall quality of research output around the world.

Though the NIH might not prioritize replication, you should. The simplest way to avoid believing in a study that turns out to be a fraud is to remember the following: No one study is definitive. No matter how incredible the results, how powerful the implications, if it has been shown only once it has not been proven. Lives would have been saved if those who were quick to embrace the Elgazzar ivermectin paper had waited for the second large trial of ivermectin, which showed that the drug had no effect at all.

It's Not All Bad

We need to acknowledge that fraud in medical research exists if we are to eliminate fraud in medical research. We need to actively look for the signs of fraud, particularly when small single-center studies have profound results that stand to earn a researcher substantial fame, power, or money.

Of course, by the time fraud is discovered, it is often too late. As Mark Twain purportedly said (but probably didn't), "A lie can travel halfway around the world before the truth gets its boots on." We need to remember that when these studies first come out, it is not immediately clear they are frauds. Discovering fraud takes investigation, and investigations take time. By the time the truth is revealed, the lie has frequently taken hold, and it may be impossible to put the genie back into that bottle.

Brian Deer may have shown evidence suggesting that Andrew Wakefield's "research" included falsified data and was conducted with the goal of founding a profitable company, but Wakefield is nevertheless a revered and celebrated figure in many circles and has leaned into his characterization as some sort of iconoclastic freedom fighter. In reality, he is the godfather of the modern anti-vaccination movement—one that has led to a surge in measles cases and deaths around the world, and that continues to lead to deaths during the coronavirus pandemic. Fraud echoes.

Sometimes, doctors have a feeling that a study simply doesn't smell right—and I'm hoping I've given you that same intuition now. Miracle cures and earthshaking results are rare and demand close scrutiny. When in doubt, the best thing we can all do is wait for independent replication. But I understand that waiting for more data can feel unsatisfying to a patient trying to make a decision in the moment, such as a decision about vaccinating their child. And when a fraud lines up with our own motivated reasoning—telling us that there is hard evidence that what we wanted to be true really is true—it can be almost impossible to dismiss.

The simple truth is that not all frauds will be caught. Once a researcher abandons the norms that guide medical research, they are beyond the ordinary mechanisms we have in place to detect shaky science. I continue to believe that outright frauds of the types I discussed in this chapter are very much the exception and not the rule. But even those rare exceptions contaminate the foundation of trust between patients, doctors, and researchers.

In my mind, sunlight is the best disinfectant. Release the data—reward those who release their data—and support replication of the studies that matter. Medical research is too high-stakes, too important, and too dangerous to be guided by norms alone. Demand better.

CHAPTER 10

Healing the System

AT FIRST, WE assumed the rhythmic clanking noise was coming from the subway car itself. But the noise was strange, more organic than metal, so Niamey (my then girlfriend, future wife) and I began looking for the source. We were shocked to see a young man sitting across the subway car, eyes closed, his mouth on one of those vertical poles people hang on to. His teeth were chomping away at the metal. Even by New York City standards, it was a disturbing sight. We watched, stunned, for maybe ten or twenty seconds.

"Drugs?" I whispered.

"I don't think so." Niamey, ever braver than me, got up and pulled him off the pole. His body jerked back into the seat, and it became clear what we were dealing with: a seizure. It was our first year of medical school. We didn't know it was a generalized tonic-clonic seizure. We didn't know what caused it, or how to treat it, or what the long-term neurological ramifications were. But we knew enough to hold the young man (Derek, we'd find out later) and try to keep him from hurting himself.

The seizure had abated by the time we got to the next subway stop—Ninety-Sixth Street on the 1 train. We pulled him off, laid him on the ground, and called 911. By the time the paramedics arrived, police in tow, he had regained consciousness and was able to talk to us and follow commands, though he was incredibly sleepy (the "postictal state," we'd learn in another year or so). The paramedics conducted a brief neurologic exam and offered to transport him to the nearest hospital.

Derek refused. He was fine, he said. This had happened before. It wasn't a big deal. The police and paramedics were satisfied when he signed a form saying he had declined medical care, and they exited the scene as quickly as they had arrived. The three of us were left to talk.

Derek was living with his brother, he told us, who took care of him. He'd had seizures for the past few years, but he had never seen a doctor about it. He took no medications, no supplements, and no drugs. I remember how flabbergasted I was at hearing this. I had never seen a full-blown seizure before, and it was terrifying. That someone could experience one and *not* seek medical help for it was virtually impossible for me to process.

I was naïve, of course. I had grown up in Connecticut with my mom, a teacher, and my dad, a banker. We had excellent health insurance and enough financial stability to cover deductibles when needed. Though I had no major maladies as a child, I saw medical care as not a *right* exactly, but as just another part of the fabric of life. Going to the doctor was what everyone did—getting checkups, getting shots, taking the occasional antibiotic, and having the reassuring knowledge that if we ever got *really* sick, this whole system would be there to take care of us.

The system, when I was a kid, was embodied by a friend of the family, about ten years older than my dad, named Bob Donaldson—Dr. Bob to me. I didn't realize at the time *who* he was: a professor and dean at the Yale School of Medicine, a chair of the Department of Internal Medicine. As a child (our families had been friends long before I was born), I mentally divided doctors into only two categories: the kindly sort and the scary sort. He was the kindly sort, with an easy smile and a casual demeanor.

Dr. Bob's house was built in the late 1700s. I loved exploring its bizarre nooks and crannies when my parents would head over there for a drink and dinner—its huge fireplaces designed to heat an entire floor, the basement that smelled of mildew and history. Occasionally, Dr. Bob would toss a pebble into the corner of the basement when he knew I was exploring down there, to make it more exciting. I knew it was him; I loved it.

While to me his expertise in woodworking, his avuncular demeanor, and his Swedish steel saw, with which we (by long tradition) cut down the family Christmas tree, were his most salient characteristics, I have a distinct memory of one medical opinion from the good doctor, because it rattled my parents so much. It was 1984 or '85—I was around six years old—and the AIDS epidemic was in the news, but it seemed distant from the suburban household in which I lived. Dr. Bob told my folks that it would be the worst infectious disease pandemic he had ever seen.

His words were prophetic, of course. And his clear-eyed epidemiologic vision was not just academic. Although he was a gastroenterologist—and a famous one at that—he changed his entire career to focus on the treatment of individuals with AIDS, creating the Yale AIDS clinic in the early 1990s. To this day, the HIV medicine service at Yale New Haven Hospital is called the Donaldson Service, named after the stomach doctor who transformed his professional life to treat the people who needed it most. I am sad that Dr. Bob never got to see me join the faculty at his old institution, see patients on the Donaldson Service, and strive to be the kind of doctor he was. He died in 2003, of complications from pulmonary fibrosis.

Dr. Bob would have taken care of Derek, and that's what Niamey and I set out to do. First, we walked him home, making sure he could get into his apartment. His brother wasn't there, but he was as safe as we could make him.

Before we left, I made a surreptitious note of the phone number of his landline, and I called the next day to check on him.

He was doing well, he said, fully recovered from the events of the night before.

I told him I'd love to get him in to see a neurologist at Columbia. I was a medical student there, I said—it shouldn't be a problem. We'd just bring him up to the clinic, get a quick opinion, and move on.

After a fair amount of cajoling, he agreed. He would take the train up to 168th Street, and Niamey and I would help him start his journey to recovery.

I stood outside the medical offices the next day, waiting for him to emerge from the subway, and was somewhat surprised when he did. After shaking his hand and thanking him for coming, I walked him into the clinic building and to reception. There, I informed the woman at the front desk that we needed to go to the neurology clinic. This individual had seizures, and I was a medical student.

She asked if we had an appointment.

"No," I said, "but, you know, I'm a student here, and we just want to head up and talk to the neurologists up there."

She looked at me like I was crazy.

Writing this now, I can hardly imagine being so green. I somehow thought that the way med school worked was that there would be a bunch of doctors hanging around, willing to see patients, use their cases to teach us, and help out in general. I hadn't really given a thought to making an appointment, or who would pay, or whether anyone needed to be paid at all. I honestly believed I could just walk this guy upstairs and get him the care he needed. That this twenty-two-year-old medical student had no idea how medical care in this country was actually delivered speaks to my sheltered upbringing.

It was the first time I realized how disconnected I was from the experiences of some of my patients.

We were told he'd need an appointment and that, to get an appointment, he'd need some form of insurance. We were handed some forms and sent on our way. That was it.

He was calm, unsurprised. I was embarrassed. It sort of all clicked at once for me. *Of course* I couldn't get him taken care of. Who did I think I was anyway? That was not how the system worked.

We said our goodbyes, and I lost track of Derek. I'm not sure whether he ever got insurance, or ever got treatment for his seizure disorder. The only thing I'm particularly sure of is that I wasted his time.

That realization—the stunningly dumb realization that not everyone in this country can get medical care when they need it—rankled me. It bothered me so much that by my third year in medical school a group of us had founded Columbia Student Medical Outreach (CoSMO), a completely free clinic for the Washington Heights community. But unlike other free clinics, specializing in urgent care or health fair–type screenings, we created the complete primary care experience. Patients came multiple times a year and were provided medications donated from various pharmacies. We secured screening mammograms and colonoscopies from the specialists at Columbia, checked cholesterol levels, and, when referrals were necessary, navigated the maze of regulations that would allow our patients to get care for free or at a tremendously reduced cost.

CoSMO was a beacon of hope in what was a dystopian nightmare of healthcare, particularly in the pre–Affordable Care Act era. But it was made possible only through charity, the generosity of literally hundreds of people volunteering time, talent, and treasure to the cause. How do you scale CoSMO to an entire country? How do you ensure that when a young man is having a seizure on the subway, he can get treatment before it happens again?

ONE OF THE greatest lies told by those with entrenched interests in our commercial, expensive, inefficient healthcare system is that it is unfixable. It is too complicated, they say, with multiple interdependent parts; changes to any one of these could bring the whole thing crashing down. It is a tactic to maintain the status quo, which benefits pharmaceutical companies, private insurers, and their stockholders but harms patients, doctors, and the trust between us.

The truth is, we all know what good healthcare looks like—it looks like a doctor and a patient working together to make the best possible decisions for the patient's health, free from external forces, bias, and agendas. It means aligning the interests of the doctor and patient so that financial incentives never dictate care. When my wife sees a patient with breast cancer and decides that she needs an MRI to further evaluate the extent of disease, she should not have to spend hours on the phone pleading her case to the patient's insurance company. When one of my patients shows up in the emergency room in a diabetic coma because he couldn't afford his insulin prescription, we should all acknowledge that the system has fundamentally failed him. We should not be satisfied that what we have is the best we can have or the best we deserve.

Physicians as Cogs in the Machine

Part of understanding how Medicine works is understanding that it works within a system. And for all of us to lead our healthiest lives, that system needs to change. We need to reshape the system to benefit doctors and patients. And I really do mean doctors and patients together. The combination of physician advocacy and patient advocacy on the same issues is a nearly unstoppable force. But for it to begin, it is important to point out how much the profession of being a doctor has changed over the past few decades.

Put simply, physicians are no longer management. We are labor. When I was young, my vision of a doctor was straight-up Norman Rockwell. The kindly, intelligent physician down the street, leather satchel in hand, who did it all. In the old model, the doctor ran his or her own practice and saw patients throughout their entire life course, delivering babies, rounding in the local hospital, attending funerals when their patients died. This model is essentially nonexistent in the United States right now. Doctors who try to do it—those who train specifically in family medicine—are a dying breed and are the second-lowest-paid doctors in all of Medicine (only pediatricians make less).

In fact, even the idea of private practice, the doctor's office as an individual enterprise, is dying. As of 2021, 70 percent of all physicians in the United States were employed by hospitals or corporations. Full disclosure: Yours truly is an employee of a corporation called Yale University. That means that we are not in charge of our own vacations, our staff, or even our own office hours. If you've wondered why your doctor has four patients scheduled for the same appointment time slot, it's likely because the company they work for set up their schedule that way. The result is longer wait times for patients, rushed appointments, and frustrated doctors.

Hospitals and corporations come with administrators. Lots of them. In 2019, *Healthline* crunched the numbers and found that there were ten healthcare administrators for every one doctor in the United States. These administrators do not provide patient care, but they supposedly *improve* care. Most administrators argue they are making the system more streamlined, more efficient. But there is not much data to support that. The *Harvard Business Review* looked at the rapid growth of healthcare administration positions and concluded that the only meaningful difference for patients that correlated with increased administrative roles was a reduction in thirty-day hospital readmission from 19 percent to 17.8 percent. *HBR* also noted that this reduction happened to occur just when the Affordable Care Act imposed penalties on hospitals based on readmission rates.

Physicians on the ground spend a fair amount of time grumbling about the administrators who dictate an ever-larger portion of our lives. Physician burnout is driven largely by requirements and responsibilities external to patient care, from dealing with insurance companies to completing the endless series of mandatory training modules handed down from the companies we work for. One such training module I was required to complete was about how best to walk in a hallway. Apparently, you should give extra space when you go around a corner to avoid bumping into someone coming from the other direction. That I needed to spend fifteen minutes of my day clicking through a presentation to this effect, and passing a quiz on the finer points, continues to rankle. I hope you'll consider this if your doctor is not in the best mood when they walk into the exam room.

The doctors' lounge at the hospital is filled with a low grumble about overadministration. The common refrain? We are the ones taking care of patients, and thus generating the income that pays administrators' salaries, so why are they running the show? This argument should sound familiar to every blue-collar worker—labor is generating capital, which is being siphoned from labor to higher-ups, administrators, and shareholders. We all need to realize that physicians have far more in common with our patients than we do with our bosses.

I'm not saying that docs need to be paid more, though I do believe some specialties like pediatrics are comparatively underpaid. Rather, I want to point out that this overadministration of

Medicine, this labor-ization of physicians, is wasteful. About 20 percent of healthcare spending in the United States goes to "physician services," but physician salaries are only a fraction of that—just 8.6 percent, the lowest percentage of any Western country save Sweden. Since the majority of physicians now work for hospital systems or corporations, that money goes to headquarters and is parceled out from there. In other words, even if you slashed doctor salaries in half, you would save only 4 percent of the Medicare budget.

Where does the rest of healthcare spending go? A large chunk goes to hospitals, which will surprise no one who has ever received a hospital bill. A recent social media trend has individuals posting their outrageous hospital bills online. A woman spent nine hours in the ER while she had a miscarriage—$8,241. A man developed appendicitis and had his appendix removed—$60,156. A man was bitten by a rattlesnake, leading to a five-day hospitalization—$153,161. The sheer audacity of these numbers defies description and explains how hospitals are taking such a large slice out of the healthcare pie.

Another slice goes to prescription and nonprescription drugs. It is difficult to know how much "administration" costs, since it pervades really all aspects of healthcare spending and can be hard to pin down. But suffice it to say, there is clearly waste in the system. How can we bring prices down?

It Doesn't Have to Be Single-Payer

You may have by now reached the conclusion that many in the United States and other countries have reached: We need universal healthcare. I agree, but I will point out that "universal healthcare" and "single-payer healthcare" are not the same thing. Universal healthcare means that everyone gets the care they need, regardless of ability to pay. There are multiple ways to deliver universal healthcare, and single-payer healthcare (where all payments for healthcare services are made by a single entity—typically the government) is just one of them.

The advantage of single-payer healthcare is that you can do away with a *lot* of administrative rigmarole. Much of what those ten administrators per doctor do is negotiate prices with insurance companies, bill insurance companies, and try to manipulate the system such that they can maximize the amount of money they get from insurance companies. With only one payer, particularly a payer that does not have a profit incentive, much of that waste could be avoided.

Politically, though, I don't see a wholesale conversion to a single-payer system happening in this country. The fear of socialized medicine (successfully practiced by basically every other functioning democracy in the world) is deep-seated in the United States. And what would happen to all those insurance companies, or, more importantly, the roughly one million employees of those insurance companies, were they made obsolete?

A viable solution to this problem is known as "all-payer rate setting," and I wish people would discuss it more. The idea here is that a hospital or healthcare provider charges the same price for the same service, regardless of who is paying. (Yes, this is how basically everything else in the world works, but healthcare loves to buck the trend.) Right now, while the "list price" of a given procedure is typically standard, hospitals negotiate discounts with every different insurance company—contracts that are hugely complicated and updated continuously. All-payer rate setting would end that. If the cost of an EKG is set at twenty-five dollars, that is the cost. All insurance companies would be required to pay that amount. Critically, patients without insurance would pay the same amount too.

Those rates could be set by an independent commission at the federal or state level. Even if no effort was made to reduce the average reimbursement for various procedures, the very fact that the *negotiation* over the prices of those procedures would be removed would reduce waste in the system. So far, one state—Maryland—has enacted an all-payer rate setting system. In fact, the state has had it in place since the early 1970s. Data suggests that the system has lowered costs for patients, and improved quality—particularly for high-expense care and for patients with high levels of medical needs. And it makes a difference to average consumers too. As of February 2022, Maryland, the seventh-most-expensive state to live in, is the third-*least*-expensive in which to buy health insurance.

All-payer rate setting also opens the door to untangling insurance status from employment status. While we have come to treat health insurance as a job benefit, it is not clear that it truly benefits the individual. The prospect of losing your insurance if you lose (or change) your job ties people into positions and industries where they may be overworked, underpaid, and unhappy. With all-payer rate setting, since there would be no need for bargaining on the costs of care, insurance could stick with the individual rather than with the job.

Making healthcare more accessible and affordable is a wildly popular idea, but it is not without its detractors. Typically, these are individuals who have excellent access to healthcare already and who worry that a level playing field has the potential to bring *their* quality of care down. For those with the means, it is reassuring to know that you can see a doctor quickly, get the testing you want, and have ready access to the most advanced therapies. In a consumer-oriented world, we are conditioned to believe that more is better. But when it comes to medical care, this may be completely wrong. If more healthcare were better, the rich and powerful would have dramatically improved health outcomes compared to the rest of us. The research shows that just the opposite is often true.

Don't Let the Chief of Surgery Pull Your Chest Tube

During my training, I was part of a team taking care of a very famous person. This individual was famous enough that, for security reasons, an entire wing of the hospital was dedicated to his care alone. Bodyguards lined the entrances and exits. He was having a relatively routine heart operation, one done hundreds of times a year at my hospital, but this situation was different. You knew it from the lines of TV trucks and cameras camped out in front of the building, but you also knew it from the demeanor of everyone on the team: quiet, nervous, excited.

The operation was a success—the height of professionalism, really. The patient was left to recuperate in his private wing, with a dedicated and discreet nursing staff to tend to bandages and two chest tubes, which were inserted (as is common practice after this type of surgery) to drain some excess fluid from the chest cavity as the patient heals. Few people were actually allowed in the room postoperatively. I was not one of them. But I was privy to the conversations the medical team had at the end of the hallway before the chief of the service would walk down that hall and enter the room to examine the patient.

On around the third postoperative day, the chest tubes were to be removed, as expected. This is a job that falls to the surgical resident, somewhere in their third to fifth year of training. They do it *all the time*. But not this time. This time, the chief of cardiothoracic surgery would be pulling the chest tube, because that's what this incredibly famous person expected. In the eyes of the incredibly famous person, who better to pull the chest tube than the chief of the entire department?

As we huddled at the end of the hallway, I'll never forget what that chief said: "I haven't pulled a goddamn chest tube in thirty years."

It went fine. There were no complications from the chest-tube pull, though I imagine the chief was sweating bullets while he did it. But the episode is emblematic of a big problem at the high end of Medicine. In fact, the problem comes with a diagnosis: VIP syndrome.

It's Not Good to Be a VIP

The term "VIP syndrome" was coined in 1964 by Walter Weintraub, who ran a small psychiatric hospital. His initial report characterized twelve individuals admitted to his hospital for psychiatric care, all of whom were VIPs (either due to their fame or power, or connections to people with fame or power). He noted the stress these individuals put on the system—including demands for expedited admissions and even free care—but what was most striking about this initial case series was the spectacular failure rate of therapy. Of the twelve VIP patients described, Weintraub points out that ten were complete therapeutic failures. Two of them committed suicide, and three left the hospital against medical advice. In other words, in Medicine, being special can be bad for you.

Papers citing cases of VIP syndrome have proliferated since the late sixties and into today. Studies have examined the outcomes of VIPs undergoing substance-use disorder treatment, ophthalmic surgery, treatment for headaches, and a host of other conditions. Broadly speaking, the outcomes have been bad. VIPs do worse. They do worse because they are not receiving the standard of care. True, in many cases they are receiving *more* than the standard of care—more tests, more interventions, more specialists, more medications. But in Medicine, more is not always better.

Several healthcare systems have what amount to institutionally supported VIP clinics. The euphemism is "executive health," and it provides a wonderful example of how having access to every possible medical service at the drop of a hat is both highly profitable for the healthcare industry and mostly useless for the patients who pay top dollar for that access.

The idea of executive health sounds appealing. Bring to bear the mighty power of diagnostic testing and physician evaluation to detect diseases in their earliest stages, such that the executives of your company—so necessary for its success—are not suddenly sidelined by something that could have been prevented. Executive health clinics conduct full-body scans of healthy individuals, do a slew of blood tests looking for cancer biomarkers, and evaluate cardiovascular function in people with no symptoms of cardiac disease.

It *sounds* good. Who wouldn't want the absolute best that healthcare can provide? But in reality, it isn't good. The reason we don't do full-body scans of every adult to detect cancer early isn't because it is too expensive; it's because it doesn't work. You are much, *much* more likely to discover an "incidentaloma"—a nonspecific something that requires more extensive testing to confirm that it is nothing—than a real disease in its earliest state.

In 2019, researchers from Washington University in St. Louis systematically reached out to executive health programs at twenty-five top-ranked hospitals and collected data on costs and what was offered. At the low end, $995 would get you a "comprehensive health assessment" at Houston Methodist. At the other end of the scale, Cleveland Clinic would enroll you in its Premier Executive Health program for a cool $25,000. The researchers identified twelve tests that would be administered to gauge cardiac health across the various programs, from stress tests to CT scans. None of the twelve tests offered are recommended by the American Heart Association to be administered indiscriminately to asymptomatic adults.

Why do these programs offer unnecessary tests? Competition. There are a limited number of individuals who can pay for (or have their company pay for) these visits, so the various programs try to distinguish themselves by offering *more*, on the assumption that the executives will conclude, like many patients do, that more is better. But for healthy individuals, there are vanishingly few examples of screening tests that prolong life or improve quality of life. You know most of them already, since these are the ones that everyone should receive whether they are a CEO or not: colon cancer screening, lung cancer screening in long-term smokers, mammograms and Pap smears for women, and blood-pressure testing.

Now, under ordinary circumstances, I'd be all for fleecing the occasional CEO. After all, the revenue brought in by executive health clinics could support multiple revenue-strapped but truly valuable programs throughout a health system (like prenatal care). But the truth is, this additional testing offers little benefit and may be harmful. In a 2014 piece in *JAMA*, Michael Rothberg related his father's experience of how one test spiraled into a nightmare of follow-up testing. His father, an older man without a smoking history but with a few minor medical conditions, saw a new primary care doctor, who conducted a thorough physical exam. The doctor noted that his aorta, the main artery that carries blood from the heart to the rest of the body, felt enlarged, so he sent him for an ultrasound. The ultrasound showed that the aorta was normal but found a suspicious-looking lesion on the head of the pancreas. A CT scan was ordered to further evaluate that lesion, and no pancreatic mass was found, but a spot on the liver was noted. A liver biopsy showed that the spot was not cancer but a benign hemangioma. However, the biopsy site bled severely, requiring a large blood transfusion.

None of these steps were unnecessary or wrong per se, but this story reminds me of the true cost of medical testing, particularly for those who are asymptomatic. As I tell my students on the wards: Be careful when you go fishing—you might catch something.

What does this have to do with patients who aren't CEOs of Fortune 500 companies? The key is understanding that the consumer mindset in healthcare can hurt you. We are all conditioned to want more for less; it is perhaps the most consistent message in society today. But when it comes to your health, more healthcare (tests, scans, even specialist visits) can harm you. The system is not designed to prevent this. Quite the opposite. Especially if you have insurance to pay these costs, there is no financial incentive for your local hospital or clinic to dissuade you from further testing or treatments.

In fact, there are only two vested interests who will speak against additional healthcare utilization. One is your insurance company, which is incentivized to get you to decrease *any* healthcare, not just unnecessary care. The other is your doctor, who, if they are a good doctor, will tell you honestly when the best thing to do is nothing.

These one-on-one conversations with your doctor are critical to the wise allocation of healthcare resources, but the reality is that most patients aren't getting their medical information in reasonable, face-to-face conversations with their physician. Increasingly, patients are getting medical information from the same source most of us get information nowadays: social media.

Rise of the Machines

The amount of medical misinformation on Facebook, Twitter, Instagram, and other social media sites could rewrite *Harrison's Textbook of Medicine* many times over. Over the course of just a few days of social media browsing, I found posts extolling the virtues of consuming *exclusively* raw meat, that cancerous tumors are a consequence of tumor biopsies, that a pH-neutral diet cures cancer, and that "germs do not cause disease." These statements are all demonstrably, flagrantly wrong and, worse, dangerous to your health. And yet we must admit they are each compelling in their own way. They speak to those simple "one thing" solutions that can be so attractive and, as such, drive a lot of engagement.

Social media companies recognize that misinformation on their platforms is a problem, but their solution is far from ideal: fact-checking. Fact-checking sounds like a good idea—under the hypothesis that those who see a compelling but inaccurate post might be equally compelled by the accurate fact-check appearing below it. But this is not the case.

First, fact-checks are not immediate; posts that are not *yet* labeled as misinformation are sometimes widely shared because people assume that the absence of a fact-check is a mark of accuracy. Few people will check back later for an update.

Second, research has shown that the presence of one of these misinformation flags actually *increases* engagement (clicks, likes, and shares) among people prone to believe misinformation. These individuals believe that "you get the most flak over the target." In other words, if the powers that be have labeled it misinformation, it must be true. The more the "shadowy cabal" fact-checks, they think, the more accurate the disputed claim is.

What other options are there? Do you censor misinformation? The issue is fraught—after all, who decides? As a defender of free speech, I shudder at the idea of certain lines of thinking being consigned to the shadows because they are unpopular. And I do not necessarily trust that fact-checkers will forever be impartial and unbiased. We may all end up on the wrong side of a fact-check in the future. If, writing prior to 1984, you'd suggested that gastric ulcers were caused by a bacteria rather than psychological stress, a fact-check would have labeled you incorrect. It was in that year that doctors Barry Marshall and Robin Warren published their finding that *Helicobacter pylori* was the true cause of gastric ulcers (and the reason they are treated with antibiotics today). They were awarded the Nobel Prize for this discovery in 2005.

Censorship is not the key to fixing the medical misinformation ecosystem (which goes hand in hand with all the other misinformation out there). The key is to change the algorithms.

The year 2021 saw a treasure trove of documents leaked from Facebook that showed, in detail we had never seen before, how the company's algorithms worked, and what type of content they promoted. Put simply, the algorithms reward outrage, and this is deliberate. In 2016, Facebook expanded the options someone has when reacting to a post. Previously, there was just the simple (some would say iconic) "like button." To that were added emojis representing sadness, love, laughter, surprise, and anger. These emojis were weighted five times as highly as a standard "like"—meaning that posts that elicited emotion were much more likely to be shown to other users, further amplifying their reach. Facebook's own data, according to leaked files, showed that posts with more *negative* reactions and comments were spread further. And as the social media "influencer" market grew, individuals became quite adept at creating content that would elicit an emotional response.

If all you care about is keeping eyes on your website as long as possible, this approach works beautifully. We can doom-scroll, we can rage-scroll, as long as we scroll. But it hurts us, sowing division, creating anger, creating a false us-versus-them narrative, negatively impacting our politics and our social cohesion. As a doctor, I also see it hurting my patients' health. It hurts them in direct ways—by promoting therapies that are at best elaborate placebos and at worst actively harmful. And it hurts their mental health—by leading people to believe the world is a hostile place, full of inhuman monsters who want nothing more than to take advantage of them, and that some of these monsters wear white coats.

It doesn't have to be this way.

What if *you* controlled the algorithm? What if you could tweak the settings on your social media feeds to say that you don't want to see as much inflammatory information or that you want to see more? That you would like to see factual posts with less emotional valence or less factual posts with more emotional valence? Implementing that functionality, with modern data analysis techniques, would be trivial. You could have the experience you want. But social media companies don't want you to have it.

I think Medicine is our "in" here. Political misinformation is a loaded issue, even when the facts are clear. But medical misinformation does obvious, direct harm. We could try to use examples of medical harm from these online communities to convince the keepers of the algorithms to start weighting fact at least as highly as emotion when they decide what to surface on your feed. It could be a beginning.

What We Fight for, Together

The fact is, doctors are aligned with patients on the vast majority of healthcare issues. Most of us want exactly the same thing you want. But the people who stand to lose when we improve the healthcare system are the people who benefit when patients and doctors are divided. We don't need to be.

We need to fight for healthcare for all. We don't necessarily need to abolish the insurance companies (all-payer rate setting, anyone?), but we need to make sure that human beings can get care when they need care. This is a moral stance, certainly, but it's also economically rational. Catching preventable diseases early saves money (provided the testing is appropriate). A 2010 article in the journal *Health Affairs* ran the numbers. If we broadly adopted twenty preventive health services (like alcohol abuse screening, and smoking cessation counseling), the United States would *save* $3.7 billion in annual healthcare costs, while saving two million life years in the process.

The Affordable Care Act protects (currently) twenty-two preventive services, ensuring that they are provided without copay or deductible. But this works only if you have health insurance, of course. Many have advocated for a "public option," a government healthcare plan (like Medicare) that could compete with private insurance or automatically enroll people who have not elected coverage elsewhere. Insurance companies are naturally averse to a public option, as the federal government has no obligation to make a profit or even break even on health insurance. But I will note that all-payer rate setting would apply here as well—leveling the playing field to allow those private insurers to compete on service if they wish.

We need to fight to make insurance more affordable. Decreasing administrative burden by eliminating the endless contract negotiations between insurers and healthcare providers would go a long way here. (See Maryland's very affordable insurance plans as an example.) We can also decrease the cost of health insurance by doing *less* Medicine—ordering fewer inappropriate tests and doing fewer unnecessary procedures. This is admittedly a delicate path to tread. While the data consistently shows that we do too much testing and operating in this country, we must acknowledge the disparities (historic and ongoing) that have led to less treatment (including less *appropriate* treatment) in marginalized populations.

We can address that, in part, by giving the time back to doctors and patients to understand *why* a test isn't indicated or a procedure won't help. Mistrust is expensive—it leads doctors to order more tests to cover their proverbial asses, and patients to find new doctors if they feel like they aren't getting all the services they want. By building trust, we can tell our patients that sometimes the best thing to do is nothing, and they may actually believe us.

We need to fight to teach science and rational thinking early and often. This will have benefits that range well beyond the medical field. To some extent, this is happening already. Kids these days are much savvier about the information they consume than they were in previous generations. Growing up with the internet means being skeptical all the time, and that will lead to better thinking later in life. But this education could be more formalized. I would love to see the return of formal logic classes as well as statistical training beginning as early as elementary school. It can definitely be done. Kids have not yet developed the deep (and often wrong) intuitions about numbers that adults have. Playing games of chance with them—rolling dice, playing cards—or measuring the odds of finding a blue Skittle in a pack of Skittles can reinforce a sense of how statistics work and can help keep them from being hoodwinked in the future.

More than that, we need to teach statistics and rational thinking in medical school. My nonmedical friends are sort of shocked to hear this doesn't happen already. To them, it's like hearing that law school doesn't discuss property rights. But traditionally, med schools do not focus on this critical type of training. They focus on the nuts and bolts: physiologic and pathophysiologic mechanisms; the way that drugs work, from a biochemical basis; the current recommended treatments for a variety of ailments. Some schools offer an Appraisal of the Medical Literature course—typically for less than a week, and often optional—but many do not at all.

What this means is that many doctors—*most* doctors, in fact—have not been trained to know which medical research is truly practice-changing. That job is outsourced to medical societies and so-called "key opinion leaders," who make guidelines and assess whether those guidelines are being broadly followed. To be clear, there is a strong evidence-based component to this practice, but in the end, it requires most physicians to trust that the American Heart Association or the American Cancer Society is doing its due diligence. Usually, these organizations are; the model works fine. But just as society at large has developed increased skepticism about institutional expertise and knowledge, so have physicians. If they are skeptical of the recommendations from the societies but *also* have no formal training on how to interpret medical research themselves, they are just as prone to be sucked into the morass of misinformation as any other individual.

We need to teach young doctors how to read a medical paper deeply and with insight, how to recognize signs of data manipulation and fraud, and how to communicate these results to their patients. Doing this across every medical school in the country would dramatically reduce the amount of medical misinformation coming from doctors themselves, which would increase trust in the entire system. Of course, bad-faith docs may promote gobbledygook to sell their own supplements or tonics, but they are easier to ferret out—they never seem quite as earnest as the docs who promote misinformation in good faith, through misunderstanding. And the fact that many bad-faith doctors sell the solution to the problem they are discussing on their personal websites is more than a little suspect.

Moving Forward

There is something so joyful about a one-on-one interaction with a patient. It is refreshing and energizing and necessary. For patients too, these visits (even though they are becoming briefer) remind us what Medicine is all about. But so much of healthcare today is not about these visits.

Everything I've discussed here is in service of making Medicine about that series of moments between the doctor and patient. Every part of healthcare that is *not* part of these moments—the insurance, the billing, the pharmaceutical marketing, the alternative therapies promoted via social media—is detrimental to healthcare.

The good news is, there are two large, powerful groups of people who (even if they all don't know it yet) are on the same side in this fight: doctors and patients. Or perhaps I should put it in language that will elicit more of an emotional response: It's us versus them.

CHAPTER 11

Moving Together

YOU CAN ASK her about the cancer, or how she's feeling, or about herself, or whatever. Just spend some time."

It was my first year of medical school, and I knew nothing. And yet here I was, at Calvary Hospital, an inpatient hospice facility in Brooklyn. My supervising physician, a palliative care doctor, had assigned me Ms. Ewing, a fifty-year-old with widely metastatic breast cancer, who had been in hospice for a little over ten days. With those words, my supervisor sent me on my way to have one of my first direct patient interactions. I was not required to draw blood, take a blood pressure, or even conduct a formal medical history. Really, I was just supposed to visit with some patients who were dying. In so doing, I would learn about the dying process, and they... Well, I suppose I hoped I would make things easier on them.

I remember thinking at first that hospice must be boring, and that passing the time, even with a stranger in a short white coat, would be better than nothing. But my brain hitched at the thought: *Pass the time* to what? Every single patient, or near enough, who came to Calvary would die at Calvary; it specialized exclusively in hospice care, after all. Ms. Ewing was here so she could be comfortable when she died. It was no more or less complicated than that.

Calvary may have been the best-run hospital I have ever seen. The hallways and rooms were immaculate, bright, and airy. The staff took their jobs seriously, but with a gentle grace that was instantly recognizable. When anyone—doctor, nurse, housekeeping staff—entered a patient's room, they smiled and said hello. They asked how the patient was feeling. If a patient wanted water, they would stop whatever they were doing and get the patient water. If you're ever in the hospital and ask your doctor for a glass of water *and they stop and get you one,* you have a special doctor. I was blown away.

I was also terrified. I knew what cancer was, of course. And I knew what hospice was, at least in theory. I knew that in this place, the goals of care were different. We weren't trying to cure; we were just trying to comfort. It freed us in many ways. Blood draws hurt, so we didn't do blood draws. When a patient was in pain, we gave pain medication. When a patient was short of breath, we gave oxygen. There were no dietary restrictions, no collection of urine and stool for analysis, no beeping IV poles, no cardiac monitors. And when a patient died, they simply died. No rapid response. No code blue. Death with a period, not an exclamation point.

Still, as I stood outside Ms. Ewing's door, knowing that my task was just to talk to her, I felt paralyzed. I didn't know where to begin. For the first time in my medical career, but certainly not the last, I opened the door without a plan.

She was pretty. She lay in bed, propped up on three pillows, a book in her lap, with her reading glasses on. Her head was wrapped in a scarf—hair loss from the chemotherapy that she would no longer be receiving—and her skin, without makeup or adornment, was nevertheless luminous. Not pale. She didn't "look like death," as the saying goes. She looked like a woman, reading in bed. She smiled, and I thought, *Is this what it is like if you know you are going to die?*

Well, what else is there? We tell ourselves to live each day as if it is our last. Maybe the converse works as well: When you're on your last days, live as you always have.

We chatted. That was it. I made sure to get her a glass of water when she asked for one. I told her about myself, about my year as a singing waiter in a restaurant in Paris. She told me about her job: a marketing executive, before. Her kids: two, in college. Her husband: coming later.

But about twenty minutes in, I froze. It was like my brain had been wiped with an eraser. I couldn't remember what we had talked about or what I wanted to ask her next. It felt like a discussion of her medical condition would be horribly out of place at this point, so I did the only thing I could think of to do. I checked my watch, told her it was nice talking to her, and left.

There was still an hour and a half before I was supposed to meet with my supervisor to discuss the experience. Ninety minutes left. I hid in a back stairwell. I sat and thought about Ms. Ewing, and how I was a terrible medical student, and how—despite being a singing waiter—I am really an introvert and don't do well being thrust into situations like that with people I don't know, and how I would die someday and so would everyone else, but that it seemed so much *worse* to know it was happening soon. Or perhaps it wasn't. Perhaps, if you were going to die anyway, you'd rather know. At least then you could set things to rights. And what would I set to rights if I had to? And what if I couldn't even if I wanted to?

"We had a nice talk," I told my supervisor, who, thankfully, didn't ask me too many more questions about it. I resolved, internally, to do better the next day. I'd spend *more* time with Ms. Ewing. I'd have topics. It would be easier.

The next day, Ms. Ewing was not available. I never found out why. She may have been feeling sick. She may have told them that she didn't much feel like talking to the awkward med student. I never got to speak with her again, actually; she died a few days later. I added an item to the list of things I'd never be able to set to rights.

Midway through my time at Calvary, my preceptor decided to observe me with a patient. An older man with pancreatic cancer, Mr. Lane was bearded and had clearly been on the heavier side before his diagnosis, but his sunken temples and bony hands bespoke the catabolic nature of cancer—the wasting disease. Affable enough, he was happy to talk with me while my preceptor sat on the other side of the room, trying to be as unobtrusive as possible.

It went well. Having been instructed to get some more details about his medical history, I had a path to follow in our conversation. The initial symptoms, the diagnosis, the treatments, where he was now, and what his goals were. The conversation flowed smoothly, and we took some detours to discuss the parts of our lives that made us feel more like real humans: our families, prior jobs, mutual love of New York–style pizza.

Eventually, as Mr. Lane was getting a bit tired, we wrapped things up. I thanked him for his time, wished him the best, and left with a "See you later."

"Will you see him later?" These were the first words out of my preceptor's mouth when we left the room.

"I . . . um . . . I didn't really mean anything by it."

"Patients listen to *every word you say*," she said. "If you say 'See you later,' he'll expect that you'll see him later. We don't lie to patients."

This was difficult feedback for the young go-getter I was at the time. Medical school had just started, and already I felt like maybe I didn't have what it takes. Sure, I could memorize the names of muscles and pills and cellular receptors, but to talk to patients and really connect with them would require a whole new mindset. It was a tough realization.

The doctor-patient relationship is unlike any other relationship in the world. For one, there is a level of intimacy that goes beyond what people have even with many of their closest loved ones. At the same time, we are not on equal ground. The doctor comes armed with years of training, with more facts and information, and with access to a prescription pad. With that pad comes the promise, often the *only* promise, that the ailment can be taken away. The doctor is somehow scientist, priest, and shaman all in one. And yet to function properly, we must somehow transcend the wall of that knowledge, that promise and hope. It would take me years before I felt comfortable doing it, and, truth be told, I'm still learning.

I did see Mr. Lane later, by the way. I made sure of it. And when I left his room that time, for the last time, I said how wonderful it had been to meet him. And I said goodbye.

IN THE LAST chapter, I wrote about the healthcare system, that flawed behemoth that can lead patients and doctors over the edge of frustration into anger and despair. My hope is that by fixing the system, we can allow the real power of Medicine—the power that exists in the doctor-patient relationship—to finally be realized. Here, I want to examine that relationship more deeply, because no relationship is perfect. Or, better, every relationship can be improved. And with the complexity of medical science as it exists right now, the strength of the doctor-patient relationship is more important than ever. To keep it strong, we need to move toward each other in several areas that don't often form the bulk of a discussion between doctors and patients. We need to focus on more than pills.

The Most Neglected Part of Health

In chapter 8, I pointed out Steven Schroeder's study that suggested that even with a perfect medical system, where everyone got (and took!) the appropriate medications for their ailments, only 10 percent of premature deaths would be prevented. But healthcare and medicine have been so intertwined that the very word "Medicine" has become synonymous with healthcare. I practice Medicine. I also prescribe medicine. But there is so much more than medicine in Medicine.

If you ask your physician what the most important thing you can do to improve your health is, they will tell you to improve your lifestyle: Exercise more, eat healthier, sleep regularly, reduce stress, stop smoking. The recognition that lifestyle choices have a meaningful and powerful effect on health outcomes really began only in the latter half of the twentieth century, as infectious disease receded as the number one killer of humans, and heart disease and cancer replaced it.

But lifestyle is inadequate, in my mind, to explain the vast epidemic of so-called "deaths of despair"—suicides, drug overdoses, and deaths from alcoholic liver disease—plaguing the United States right now. Beginning around 2000, epidemiologists noticed an uptick in these particular types of deaths, but the magnitude of the problem was not fully clarified until recent years. While drug overdoses in the setting of the opioid epidemic have driven some of the increase, the last time suicide levels reached as high as they are now was in 1938, during the Great Depression. The last time deaths from alcoholic liver disease reached this level was in 1910. Overdose deaths have never even been close to this high. And while the increase in deaths of despair was first reported among white men, recent research shows similar trends among Hispanic and Black men, and women as well. Despair is becoming universal.

The data is completely clear: This *is* happening. But researchers, politicians, and pundits differ as to why. Some point to economics—increasing economic inequality and the dominance of social media can make individuals feel they will never achieve what they perceive their peers are achieving. Some point to globalization, with previously stable communities built around manufacturing falling apart. Some point out that we are among the first generations to feel we are achieving less than our parents did, the central promise of the American dream unfulfilled. Some point to the decline of marriage rates and the nuclear family.

What strings all these theories together, though, is social isolation. I was about to explain how despite our incredibly interconnected world, people can still feel isolated, but I don't have to, do I? I suspect that every single person reading this book understands how scrolling through pictures of your friends on a small phone in a dark room could make you feel completely, utterly lonely.

Loneliness has a powerful effect on health. Research demonstrates that social isolation and feelings of loneliness lead to impaired executive function, worse sleep, and declines in mental and physical well-being. A large meta-analysis, combining results from seventy studies and three million participants, found that individuals who felt lonely died at a rate roughly 30 percent higher than those who did not. This makes social isolation and loneliness significant risk factors for death. What's worse, the healthy way to cope with loneliness—reaching out to others—is not the most common way Americans deal with it. A Kaiser Family Foundation poll in 2018 found that the number one way we cope with loneliness is by distracting ourselves with TV, computers, or video games.

If loneliness is a significant risk factor for death, it would seem that your doctor should be asking about your loneliness at every visit, but we don't. It's an awkward question, to be sure, but I don't think "Are you lonely?" is particularly more loaded than "How many sexual partners have you had in the last year?" Rather, I think we don't ask for two reasons: First, we aren't sure what to do with the answer. I have no medicine for loneliness. It is not a psychological disorder; it is a social one. The other reason is that, frankly, I have another patient I need to see within the next ten minutes, and I'm afraid of opening a can of worms.

But we *need* to open that can. When I think back about my most positive experiences in Medicine, the times I felt really good being a doctor, one common theme emerges: I had time.

I had enough time to sit with a patient and talk. I was not rushing to the next exam room, or meeting, or home to feed the kids. It was when I'd spent enough time with a patient that I discovered spousal abuse. It was when I'd spent enough time with a patient that I learned of a drug use disorder they had been hiding for years. It was when I'd spent enough time with a patient that I got the recipe for Midnight Pasta, which I use to this day.

Time is necessary to make those connections. It is also sufficient. In other words, it is really all we need. There is a nearly 1:1 correlation between the time I can spend with a patient and our mutual satisfaction with that meeting.

Physicians must attend not just to the biological health of our patients but to their social health as well. We must take the time—*demand* to have the time—to uncover those issues, because they do not come out in the first five minutes of a visit. Of course, with ten administrators for each physician counting the patients we see on a daily basis, the demand for more time is usually met with a scoff or, among the more humane administrators, a sigh of understanding and a shrug, as if to say "It is a beautiful dream, and yet..."

Perhaps if physicians truly internalized that we are labor and not management, we would be more comfortable fighting for better working conditions—that is what time provides. Patients can support these efforts by expressing their concerns to hospital and health system administrators, using language administrators understand, such as "If my doctor does not have adequate time to care for me, I will need to seek out another doctor."

The first step for doctors to earn your trust back is spending time with you. We need to be allowed to do this. We need to fight for it. But it isn't the last step. Doctors also need to understand that their patients' symptoms are real and important even when they are not life-threatening.

Nonfatal Diseases Are Still Diseases

To save a life is an incredible feeling. Chasing that feeling drives many physicians who work to save individual lives every day, and many scientists who work to save thousands or even millions of lives. And yet most patients do not have a life-threatening condition. Doctors' myopic focus on lifesaving, in some sense, means a loss of focus on quality of life—on the very issues that matter most to the vast majority of our patients.

The problem is that physicians are trained to evaluate a symptom in terms of life-threatening conditions first. If someone comes to my office noting palpitations, I am trained to make sure they didn't have a heart attack, a life-threatening arrhythmia, pheochromocytoma (a rare tumor), and so on. Medical school reinforces this style of thinking, and while non-life-threatening conditions like anxiety and panic attack are on our potential diagnosis list when someone comes in with palpitations, many doctors feel the workup is essentially complete once life-threatening issues are ruled out.

This leaves patients feeling abandoned. A symptom develops, and a thoughtful physician puts together a diagnostic plan that seems reasonable. The patient follows through diligently, getting the prescribed tests and procedures. And then the patient is told, "Good news—the debilitating symptom you have is not going to kill you," and that is supposed to provide closure. But the symptom is still there. The patient goes to another doctor, often repeating the *same* tests, and getting the same "good news." The process repeats.

Doctors need to recognize this cycle and break it. Patients can help us recognize it by acknowledging their relief that the condition is not life-threatening and then simply asking "So, what is the next step?"

We don't have great tools to diagnose non-life-threatening conditions because most research focuses on the diagnosis and treatment of life-threatening conditions. We are terrible at determining what is really causing back pain, random numbness and tingling, stomach and gastrointestinal problems, and mild psychiatric issues, to name just a handful that can be life-altering for patients. In fact, we are only beginning to learn how to even measure the effect of these ailments (since "death rate" won't work).

Fortunately, there is some movement in this area. In the past decade, there has been a real shift in federal research-funding agencies to include "patient-reported outcomes" as primary targets for research studies. Patient-reported outcomes are just what they sound like—instead of a doctor or researcher measuring something in the blood or looking up data in a medical record, patients themselves tell us how the treatment made them feel. Standardized surveys, tested across thousands of patients, have been developed to assess things like the impact of pain on quality of life, the scope of depression and anxiety, the burden of symptomatic but nonlethal diseases.

This is good progress. But it isn't enough. We need to train our young doctors to focus on these issues, because focusing on what matters to patients builds trust with patients. If I can help you with the nonlethal thing, you will trust me when we need to treat the potentially lethal thing.

In my own practice, I have a habit of making this a bit transactional, honestly. If it's a regular "check-in" type of visit, I ask the patient what the number one thing that is bothering them is. It doesn't have to be their most serious disease—it's the thing that they want to talk about. I take that in and tell them what my number one issue for them is. (It may not be the same.) We then agree that we'll address both. We'll work on your back pain that makes it hard to sleep and *also* get your cholesterol under control to avoid another heart attack. We'll figure out why you get recurrent headaches at eight o'clock every night and *also* discuss a new medication that might help your kidneys function longer. Our priorities, though different, synergize if we address them together.

As doctors learn to grasp the true impact of their patients' symptoms, patients can move closer to doctors as well. The key to that step is appreciating uncertainty.

Humility *Is* Good Bedside Manner

As a kidney disease fellow at the Hospital of the University of Pennsylvania, I spent a lot of time in the cardiothoracic intensive care unit. Open-heart surgery is both a delicate and brutal affair, with patients emerging from the operative suite with all manner of lines, tubes, and medications flowing into every available vein. A miracle, to be sure—but it could also be a disaster.

Kidney damage was common. It was so common, in fact, that on nights when I was on call I would stop by the CT-ICU before I left for home. "Do you need me now?" I would say. "I'd rather see the patient now than at two a.m." Many nights, I came in at 2 a.m. anyway.

I remember listening to one of the cardiothoracic surgeons talking to a family after a large aortic arch repair. The patient had done well but was still unconscious, on the breathing machine. The surgeon, an expert by any definition, explained that the surgery was essentially flawless; in a few days, the patient would be up and around, out of the hospital within two weeks.

"Thank God," his wife said.

"No, no. You do not thank God," the surgeon said. "You thank me."

And she did.

As you can imagine, I've relayed that story many times over the years, as it encapsulates much of the stereotypes (not entirely untrue) about surgeons in general and cardiothoracic surgeons in particular. They can be . . . well, a bit cocky.

And to be honest, I think patients like that. Not the "I am better than God" thing. At least, not when stated so directly. But patients really do value certainty, confidence, and even hubris to a point.

When patients are getting second and third opinions, they often have a hard time figuring out just *why* they chose surgeon #2 over surgeon #1, but I've found that confidence usually plays the major role. When pushed, patients will say that surgeon #2 "seemed to have it all figured out" or "laid out a really clear plan" or "knows exactly what they are doing."

This is a problem. Medicine is not a certain business, as this book has repeatedly made clear. It is a science of percentages, a science of intelligent guesses, a science of hedging bets. But if you ask your doctor whether you should take metoprolol or carvedilol to treat your heart failure, which answer would you want?

Doctor #1: "The data suggests they are fairly equivalent, so whatever is less expensive is probably fine."

or

Doctor #2: "Absolutely carvedilol. I've had tremendous success with that drug."

Most people would want the answer of doctor #2. It's reassuring and confident. And (since the two drugs really are equivalent in this case) not likely to harm the patient (unless the cost their health insurer has negotiated for carvedilol is higher than the cost it negotiated for metoprolol).

But I want everyone reading this to beware of doctor #2. Doctor #2 has learned that, in Medicine, certainty is valuable, even marketable. Doctor #2 is doing what many practitioners of alternative medicine do: selling certainty.

It is no coincidence that patients consistently rate the bedside manner of acupuncturists and chiropractors higher than that of traditional physicians. Confidence is powerful. (Acupuncturists and chiropractors also spend more time with their patients.) To compete with traditional medicine, alternative practitioners need to be very, very confident. They need to make big promises, like "This treatment *will* cure your back pain." And if it doesn't, they can just move the goalposts: "This next treatment will cure your back pain." It's a model that works astonishingly well.

That isn't *real* bedside manner, however. Real bedside manner doesn't come from talking; it comes from listening. My wife, a breast surgeon, has real bedside manner. She is confident, of course, as she should be, given her years of training and experience. But that is not what her patients, when I have met them, mention. They tell me how she *understands* them. How she *listens* to them. How she is *honest* with them. That is what you need to look for in a physician.

Part of that honesty is humility. We know a lot, but there is a lot we don't know. Find yourself a physician who can admit that. A physician who says "Here is my best advice—I think it will help," not one who promises the world. Find a physician who can say "I don't know." Because, you know what? That is the truth in many circumstances. And, perhaps most importantly, find a physician who can say "I'm sorry."

The Power of Apology

The tools in the medical arsenal are powerful but dangerous. Used properly, they save lives. But mistakes happen. Apologizing to a patient for a mistake is one of the more difficult things I've ever had to do. Yes, there's the malpractice issue—some law firms advise doctors to make nonapology apologies like "I'm sorry you feel that this was the wrong treatment" or "I'm sorry you had a bad outcome," but these don't build trust between patients and doctors any more than they do between politicians and the public when they make the same kinds of statements. It's not really the fear of lawsuits that makes apologizing tough, though; it's grappling with an injury to our self-image. I didn't get into Medicine to hurt people. But, like all doctors, I have.

I'm reminded of an error—a series of errors, really—I made in the ICU back at the University of Pennsylvania. A patient, Ms. Moya, had been admitted in septic shock, comatose and intubated, and she needed powerful drugs to keep her blood pressure high enough to at least perfuse her vital organs. These drugs, like adrenaline and noradrenaline, cannot be given through a typical IV; they can damage and block off the delicate veins in the arms and hands. Rather, we insert what is called a "central line," a long catheter, typically placed into the jugular vein with its tip just above the right atrium of the heart. The blood flow there is very high, allowing quick dilution of tissue-damaging chemicals.

But by the time she came to us, her blood pressure was so low her heart had nearly stopped. I needed a central line quickly, and, at the time, the best way to do that was to place it in the femoral vein at the top of the thigh. The procedure is a bit tricky, though. The femoral artery lies right next to the femoral vein—sticking the wrong vessel can lead to rapid blood loss. Fortunately, it's pretty easy to tell if you've hit the artery. Arterial blood is bright red and pulsing. Venous blood is dark purple and flows slowly. Of course, when someone has very little blood pressure to begin with, even arterial blood may look a bit venous.

I was careful. I inserted a finder needle into the vein and saw the slow purple flow of blood come back out. I then threaded a wire through that needle, to keep my place, made a tiny nick in the skin, and passed a dilator over the wire. Blood flowed more briskly after that, but that is expected. I then took the long catheter, threaded it over the wire, sutured it into place, and bandaged the area. Life-saving medications were infused, and Ms. Moya's blood pressure started to rise.

About an hour later, the nurse called me. The site was bleeding. I took a look. There was certainly more blood than expected, but not a frightening amount. Due to the septic shock, Ms. Moya's platelets (the cells that help form blood clots) had dropped to near zero. *That's surely the reason she's having some oozing at the catheter site,* I thought. *We'll apply pressure. Nothing else to be done.*

Six hours passed with more bleeding. The dressings had to be changed several times, and I was called in to evaluate the patient again. I started to worry that I might have hit the artery, not the vein. After all, she was so sick when I'd put the line in that perhaps I didn't recognize that it was arterial blood. There was an easy way to test this—I drew some blood from the line and sent it to the lab to measure its oxygen content. At the same time, I drew blood from her radial artery and cephalic vein and sent those for oxygen levels.

The results came in: The radial artery blood was high in oxygen, the cephalic vein blood low in oxygen. And the blood from my line? Low in oxygen. It was not in the artery.

Relieved, I told the nurse that it must be the low platelets causing the oozing. We were transfusing platelets frequently, but they were getting used up as quickly as we could give them.

By the following morning, the site continued to bleed, and I was basically out of ideas. One of my colleagues suggested we get an ultrasound to check the area. I didn't know exactly what we'd be looking for, but I agreed. He probably saved Ms. Moya's life.

The ultrasound showed something I had never seen before or since, and something the radiology team hadn't seen either. The line was in the vein, as expected. But on its way to the vein, it had *passed through* the artery. The artery was bleeding, though the blood we'd withdrawn from the line was venous. "A thousand-to-one shot," they told me.

This was not something that I could fix. I called vascular surgery, who took her quickly to the operating room to repair the damage I had done. And I was left to explain to her family exactly what had happened.

Looking back, perhaps because of the shame and shock I felt at the time, I think I was too clinical, describing to her family in more detail than was necessary *why* she needed the central line, how you do the procedure, exactly what went wrong, and so on. I said that it was my fault, since it was, and that I was sorry.

They were incredibly gracious, and for that I am ever thankful. Ms. Moya recovered, I was not sued, and life went on.

There were more apologies to make in the future. And with every one, I found myself wishing that someone had taught me how to do this. Apology training is not something taught in most medical schools.

It could be, though; there is quite a bit of science behind it. Writing a commentary for *JAMA* in 2006, psychiatrist Aaron Lazare outlined the four main pieces of an effective apology: acknowledgment of the mistake, an explanation of the mistake, an expression of remorse, and addressing reparations (which can range from a commitment to a full investigation, from sharing of results to financial compensation).

An effective apology is healing in a variety of ways. It places the fault of the event on the physician, rather than the patient's family members, who often blame themselves for any adverse event that occurs to their loved one in the hospital. It validates a set of shared values: showing that what *you* think is wrong, *we* also think is wrong. If the patient was treated unkindly or disrespectfully, an apology can restore a sense of self-worth. Apologies transfer power from the physician to the patient, or the patient's family—an important component of building trust, since medical care can be profoundly disempowering. And there can be a sense of justice here. Lazare notes that some patients and families may be gratified to see that the physician who committed the error is suffering because of that error, even simply through humiliation.

But this process needs to be taught and, I daresay, protected. Physicians—particularly surgeons—have a long tradition of private honest discussions about medical errors, mistakes, and bad outcomes. These morbidity and mortality conferences are conducted behind closed doors, and they are a fascinating window into how doctors really think. At a typical M&M, as they are called, cases with bad outcomes will be presented frankly.

The goal is not to assign blame to a single person per se, but to identify the processes that allowed the harm to happen. A patient developed a severe diarrheal infection after surgery—why? A patient died due to bleeding after a kidney biopsy—why? A patient was treated with the medication prescribed to their roommate—why? While human error is often a part of these discussions, blaming human error alone is completely counterproductive. Systems need to be in place to identify and catch human errors before they reach the level where they might harm patients, because, after all, we are only human. We *all* make mistakes.

M&M conferences are real, raw Medicine, and something that

patients never get to see. The private nature of them allows physicians to ask one another tough questions. It also allows us to support one another, to remind one another that we have all "been there."

I'm not saying we should invite patients to M&M conferences. Though I'm not strongly opposed to the idea, I think the presence of someone who suffered due to a medical error would limit the openness of the discussion. But I do think that physicians should take the spirit of the M&M conferences back to the families—that same level of introspection, that same level of honesty. Virtually all surgical training programs have frequent M&M conferences. Fewer medical training programs do, but they really are crucial.

It bears noting that M&M conferences are, in general, protected under law. Notes and transcripts are considered inadmissible when lawsuits are filed on behalf of injured patients, because of the recognized value that M&Ms bring to medical care as a whole. Currently, many states have some version of "apology laws" on the books, which say that expressions of regret are not admissible in court, but acknowledgment of fault may be. These laws are rather weak and probably haven't had the desired effect of reducing malpractice claims. A 2019 study from the Agency for Healthcare Research and Quality showed that after introduction of apology laws, malpractice cases actually increased, and award amounts did as well.

In my mind, this finding is neither here nor there. Malpractice claims are costly, but they are not the predominant driver of medical costs, and, frankly, when someone is harmed, it seems reasonable that they should be compensated. Suing a doctor does not have to be punitive, though it certainly feels that way to the doctor. The best way to reduce malpractice claims is to reduce malpractice. And the best way to do that is to improve the systems that support medical professionals as they try to deliver the best quality care.

The Path to Trusting Medicine Again

In deciding to write this book, it was my hope that being honest—sometimes brutally honest—about Medicine would be the first step toward healing a rift that has been growing between the public and doctors over the past several decades. Some people told me it would have the opposite effect, that exposing the limitations of Medicine and medical science would make people trust their doctor less, not more. (I devoted a whole chapter to discussing how medicines might not work, after all.)

But I believe that people are intelligent enough to embrace the central contradiction in medical science: that an enterprise that is often focused on profits over people, that largely ignores huge problems like social isolation and non-life-threatening illnesses, can still be a force for good, that it can still be the single endeavor that has alleviated more suffering in humankind than any other. In other words, look where we are, but imagine where we could be.

Getting to where we could be requires a lot of change. I have outlined some changes we could make as a society to how healthcare is delivered, but this is not a health policy book. This is a book to help us all think about Medicine: what it really is, and how it can help us.

For patients, this book asks you to think about Medicine in a new way—less like a miracle pill and more like a stock investment. Doctors help you invest in your health, but there are no sure things. The book also asks you to recognize how powerful the *idea* of a sure thing is, and how other people—some with good intentions and some with bad—can use that idea to take advantage of you when you are in a time of need. This book asks you to appreciate how Medicine continues to evolve, with new discoveries overturning old beliefs, inching us closer to real truth. It asks you to recognize your own human biases, to avoid believing something because you *want* it to be true or because it comes from someone you want to believe.

This book demands something of physicians as well. Fundamentally, I want this book to force doctors to ask a simple question: "Whose side am I on?" And to realize, I hope, that there is no difference between being out for ourselves and out for our patients—because doctors and patients are on the same side.

It is doctors and patients who fight together against disease, death, pain, and suffering. It is doctors and patients who fight to improve the human condition, to create the world we live in today—a world of breakthroughs and a world of cures. With those battles under our belt, there is no reason we can't fight together to ensure that every human being gets cared for when they are sick, or that life-saving medications are affordable, or that mental and social health are as venerated, researched, and supported as physical health, or that patients get the time with their physicians to ask all of their questions, without worrying about the cost. We can do this. And, really, we must.

If we don't—if we continue to turn Medicine into a machine, albeit a powerful one—we will continue to see a breakdown in the trust of doctors and science in general. We will not be treading water; we will be sinking. Mistrust will lead the public away from safe, proven medical therapies, leading to further declines in life expectancy. Deepening opposition to public health interventions will lead to worse disease outbreaks, as we saw with COVID-19. Trust will be based on charisma rather than facts, deepening divisions as people choose sides without evidence.

I know the problems extend beyond Medicine. We live in an era where fact and truth seem to have little to do with each other, where people with real expertise are lambasted and people who profess to be experts are extolled. We live in an era where, as William Butler Yeats so eloquently put it, "the best lack all conviction, while the worst are full of passionate intensity." Is my belief that restoring trust in Medicine might be the beginning of a larger restoration of trust in society naïve? Is it my own bias manifesting "When you're a hammer, all the world's a nail"? Perhaps. But I hope not.

Because Medicine isn't politics. It isn't faith. It isn't tribal or partisan. It has been with us from the beginning—our deep ancestors trying, failing, and trying again to heal one another. It is the oldest form of compassion, the oldest form of empathy. The scientific method changed Medicine forever, making it more powerful, more perfect, and more valuable. But the gesture of a doctor placing a hand on the hand of a patient is the same now as it was fifty, five hundred, and five thousand years ago. It is no less powerful and no less sublime now than it was then.

Hand in hand, we can fulfill the promise of Medicine, and maybe heal more than one another.

Acknowledgments

NONE OF THIS—this book, my job, modern medical science—would be possible without the individuals who volunteer to participate in medical studies. They took a risk so that the rest of us could benefit. That is true altruism. We are moving closer to a cure together.

The doctor I am today was shaped by all the patients I've cared for (and who have cared for me) as I learned, and continue to learn, this humbling profession. I am ever grateful that the path of my life has intertwined, however briefly, with theirs. While I was able to write about some of my patients in this book, there are many more who have had an influence on the way I practice Medicine and the research questions I want to ask. Thank you for letting me fight with you.

I would also like to thank the amazing teachers who inspired me throughout my life—to question, to doubt, to experiment, to try to understand, and, eventually, to start teaching myself. I am grateful for Linda Carleton, whose comparative religion course in high school opened my eyes to the idea that people can understand the universe in so many wonderful ways. And for Bob Vance, whose infectious enthusiasm for science (and willingness to let me experiment on my own) has stuck with me ever since freshman physics. My adult interests were particularly shaped by Don Landry at Columbia College of Physicians and Surgeons, who taught me that understanding something deeply is so much more satisfying than being mystified by it.

Mentorship is so critical in research, and I benefited from having some terrific mentors. In particular, I'd like to thank Harv Feldman, whose casual epidemiologic brilliance has always been aspirational. It's Harv's voice that is still in my head asking me if I truly *believe* my findings. Chirag Parikh helped me transition from mentee to mentor and taught me how to run a lab—though my feet still don't fit in those big shoes.

Writing a book is hard. Despite having written hundreds of research papers and articles before embarking on this endeavor, the challenge of creating something cohesive around so many ideas was too much for me alone. To that end, I want to thank my agent Howard Yoon, who believed in this idea way back when I thought we'd spend more time talking about propensity scoring and causal inference, and who stuck with it until it became something I am truly proud of. Thanks to Eliza Barclay of Vox.com, who suggested I reach out to Howard in the first place. Thanks to Colin Dickerman, my editor at Grand Central, who has never had a bad idea, and reminded me to remember who this book was really for. Also thanks to Haley Weaver, who was able to see things in this work that I couldn't. Your perspective and inclusiveness reminded me how much the system we've created can hurt those who don't have the same privilege I do. Elizabeth Johnson is a simply outstanding copy editor and fact-checker, who is singlehandedly responsible for saving me from multiple embarrassing gaffes that have proved my memory for facts is not quite as reliable as the primary literature. Thanks to my sister Lee-Ann Harris and my parents Pam and Buck

Wilson for their critical and thoughtful eyes on a late draft, and to Meredith and Luke Rocklin, who have always been my biggest cheerleaders and best family friends.

I spend most of my days at Yale's Clinical and Translational Research Accelerator with members of my work family, who have heard about this book for the past two years in one form or another. I am thankful for their indulgence as I tried to get this project over the finish line. The team—researchers, statisticians, data scientists, technicians, students, coordinators, administrators—are all consummate scientific professionals, but they are also wonderful, compassionate individuals. They are the very engine of progress. Thanks especially to Deb Kearns, who somehow managed to keep my schedule intact during this period. Butterfly to the moon.

I'd like to offer a brief word of acknowledgment to all the young researchers out there who feel disenchanted, daunted, and disheartened at the prospect of starting an independent research career of their own. I hear you. When I was in your shoes, there was a program that would bring these successful researchers in to talk to us young'uns, theoretically to inspire us. We'd ask Dr. Award Winner how they got where they got and, inevitably, they would discuss some serendipitous finding that opened the door to this really productive new area of discovery. At the time, I worried this was all selection bias. They weren't going to bring all the failed researchers in to give us inspirational talks. Was enjoying a successful research career just luck? Choose the right thing to study at the right time, and if you don't, you fail? The truth is, it is a bit about luck. When your first grant gets reviewed, it's luck whether the reviewer just had a filling lunch, or a fight with their spouse, or read the world's greatest grant five minutes ago. When your first paper is submitted, there's some luck as to whether the editorial board member is feeling gracious enough to send the thing out for peer review. What do you do when hitting the goal is a matter of luck? You take a lot of shots. You increase your odds of getting lucky. Don't be discouraged by the first miss, or the second, or the tenth. A friend once told me, "If you can't control the outcome, you better love the process." And that's the only real advice I have: If you love taking those shots, keep taking them.

My exposure to medical research has been broadened substantially by the team members at Medscape—who are constantly feeding me new studies to analyze and comment on. Thanks to Eugenia Yun, Laura Stokowski, Madeline Farber, and Christine Wiebe for all your help on Impact Factor and the Medical Minute. Thanks to Roger Sergel, who has been with me since my first video commentary, as well. No science communicator is an island. Colleen Moriarty at Yale has been instrumental in connecting me to the news organizations that want to responsibly report on medical issues. Matthew Reynolds and Doug Forbush did amazing work putting together our Coursera course. And a deep thanks to Ryan McEvoy, who somehow manages to get me into the studio and film me week after week despite a million other responsibilities.

I'm not sure what I'd do without Beth Schaffer, who loves my kids like Niamey and I do and keeps them safe, happy, and fed while I am at the office, in the lab, or in the hospital. When the workday is done, I come home to my wife and kids, who are my refuge. Elaia, Huxley, and Ina can make me laugh when I need it most—and I needed it a lot over the past two years. They were also the most vicious title editors I have ever seen, never holding back when a proposed title was "terrible," "crazy," or worthy of an epic preteen eye roll. And my deepest thanks to my wife, Niamey—the best doctor I know—and also my best friend, my best editor, two-time long drive champion, and the only person who consistently laughs at my terrible jokes. Without her support, I would have given up on this long ago.

Notes

INTRODUCTION

x **That's how long the typical doctor waits:** Naykky Singh Ospina et al., "Eliciting the Patient's Agenda—Secondary Analysis of Recorded Clinical Encounters," *Journal of General Internal Medicine* 34, no. 1 (January 1, 2019): 36–40, https://doi.org/10.1007/s11606-018-4540-5.

xiv **Combine our limited schedules with a seemingly:** Melissa Bailey, "Ambulance Trips Can Leave You with Surprising—and Very Expensive—Bills," *Washington Post*, November 20, 2017, https://www.washingtonpost.com/national/health-science/ambulance-trips-can-leave-you-with-surprising--and-very-expensive--bills/2017/11/17/6be9280e-c313-11e7-84bc-5e285c7f4512_story.html.

xiv **It is no wonder why:** Robert J. Blendon, John M. Benson, and Joachim O. Hero, "Public Trust in Physicians—U.S. Medicine in International Perspective," *New England Journal of Medicine* 371, no. 17 (October 23, 2014): 1570–72, https://doi.org/10.1056/NEJMp1407373.

xix **It's no mystery why before the modern era:** "Mortality in the Past—Around Half Died as Children," Our World in Data, accessed June 13, 2022, https://ourworldindata.org/child-mortality-in-the-past.

xix **Ninety-five percent of humans:** Ibid.

CHAPTER 1: OUR MOST HUMAN FAILING

3 **Papers out of China suggested:** Benjamin H. S. Lau, Padma P. Tadi, and Jeffrey M. Tosk, "Allium Sativum (Garlic) and Cancer Prevention," *Nutrition Research* 10, no. 8 (August 1, 1990): 937–48.

3 **American papers linked eating:** Khalid Rahman and Gordon M. Lowe, "Garlic and Cardiovascular Disease: A Critical Review," *Journal of Nutrition* 136, no. 3 (March 1, 2006): 736S–740S, https://doi.org/10.1093/jn/136.3.736S.

7 **But even putting ethics aside:** Monica Dinu et al., "Vegetarian, Vegan Diets and Multiple Health Outcomes: A Systematic Review with Meta-Analysis of Observational Studies," *Critical Reviews in Food Science and Nutrition* 57, no. 17 (November 22, 2017): 3640–49, https://doi.org/10.1080/10408398.2016.1138447.

8 **One of my favorite quantitative:** Peter H. Ditto and David F. Lopez, "Motivated Skepticism: Use of Differential Decision Criteria for Preferred and Nonpreferred Conclusions," *Journal of Personality and Social Psychology* 63, no. 4 (1992): 568–84, https://doi.org/10.1037/0022-3514.63.4.568.

10 **And interpretation, if you let it:** Arthur L. Kellermann et al., "Gun Ownership as a Risk Factor for Homicide in the Home," *New England Journal of Medicine* 329, no. 15 (October 7, 1993): 1084–91, https://doi.org/10.1056/NEJM199310073291506.

12 **It was in 2003 that mRNA vaccines hit their stride:** Yen-Der Li et al., "Coronavirus Vaccine Development: From SARS and MERS to COVID-19," *Journal of Biomedical Science* 27, no. 1 (December 20, 2020): 104, https://doi.org/10.1186/s12929-020-00695-2.

12 **But the facts most commonly misinterpreted:** Vaccine Adverse Event Reporting System (VAERS), accessed February 16, 2022, https://vaers.hhs.gov.

12 **This has resulted in some humorous:** Harriet Hall, "Reality Is the Best Medicine: Diving into the VAERS Dumpster," *Skeptical Inquirer*, November/December 2018, https://skepticalinquirer.org/2018/11/diving-into-the-vaers-dumpster-fake-news-about-vaccine-injuries.

13 **As of June 13, 2022**: United States Department of Health and Human Services (DHHS), Public Health Service (PHS), Centers for Disease Control (CDC) / Food and Drug Administration (FDA), Vaccine Adverse Event Reporting System (VAERS) 1990–06/03/2022, CDC WONDER Online Database, accessed at http://wonder.cdc.gov/vaers.html on June, 13, 2022, 1:12:10 PM.

13 **But the political divide with regard to:** Ashley Kirzinger et al., "KFF COVID-19 Vaccine Monitor: July 2021," Kaiser Family Foundation, August 4, 2021, https://www.kff.org/coronavirus-covid-19/poll-finding/kff-covid-19-vaccine-monitor-july-2021.

14 **As of this writing, these approaches:** Matthew E. Oster et al., "Myocarditis Cases Reported After mRNA-Based COVID-19 Vaccination in the US from December 2020 to August 2021," *JAMA* 327, no. 4 (January 25, 2022): 331–40, https://doi.org/10.1001/jama.2021.24110.

14 **Compared to the risk of death from COVID-19:** Anirban Basu, "Estimating the Infection Fatality Rate Among Symptomatic COVID-19 Cases in the United States," *Health Affairs* 39, no. 7 (July 2020): 1229–36, https://doi.org/10.1377/hlthaff.2020.00455.

16 **Doctors' predictions were almost always:** E. Chow et al., "How Accurate Are Physicians' Clinical Predictions of Survival and the Available Prognostic Tools in Estimating Survival Times in Terminally Ill Cancer Patients? A Systematic Review," *Clinical Oncology* 13, no. 3 (June 1, 2001): 209–18, https://doi.org/10.1053/clon.2001.9256.

16 **One powerful tool that has emerged in this:** Alvin H. Moss et al., "Prognostic Significance of the 'Surprise' Question in Cancer Patients," *Journal of Palliative Medicine* 13, no. 7 (July 2010): 837–40, https://doi.org/10.1089/jpm.2010.0018.

18 **In May 2016, the results:** Alexander Zarbock et al., "Effect of Early vs Delayed Initiation of Renal Replacement Therapy on Mortality in Critically Ill Patients with Acute Kidney Injury: The ELAIN Randomized Clinical Trial," *JAMA* 315, no. 20 (May 24, 2016): 2190–99, https://doi.org/10.1001/jama.2016.5828.

18 **The conclusion: An early start:** Stéphane Gaudry et al., "Initiation Strategies for Renal-Replacement Therapy in the Intensive Care Unit," *New England Journal of Medicine* 375, no. 2 (July 14, 2016): 122–33, https://doi.org/10.1056/NEJMoa1603017.

CHAPTER 2: CHANGING OUR MINDS

24 **critical care specialist Emanuel Rivers:** Emanuel Rivers et al., "Early Goal-Directed Therapy in the Treatment of Severe Sepsis and Septic Shock," *New England Journal of Medicine* 345, no. 19 (November 8, 2001): 1368–77, https://doi.org/10.1056/NEJMoa010307.

26 **In 2017, sixteen years after the Rivers:** Protocolized Resuscitation in Sepsis Meta-analysis (PRISM) Investigators, "Early, Goal-Directed Therapy for Septic Shock—a Patient-Level Meta-Analysis," *New England Journal of Medicine* 376, no. 23 (June 8, 2017): 2223–34, https://doi.org/10.1056/NEJMoa1701380.

29 **In 2018, psychology professor Kristin Laurin published:** Kristin Laurin, "Inaugurating Rationalization: Three Field Studies Find Increased Rationalization When Anticipated Realities Become Current," *Psychological Science* 29, no. 4 (April 1, 2018): 483–95, https://doi.org/10.1177/0956797617738814.

30 **First described in 1977, the effect:** Lynn Hasher, David Goldstein, and Thomas Toppino, "Frequency and the Conference of Referential Validity," *Journal of Verbal Learning and Verbal Behavior* 16, no. 1 (February 1, 1977): 107–12, https://doi.org/10.1016/S0022-5371(77)80012-1.

30 **From April 2016 to October 2020:** Search on Trump Twitter Archive, accessed June 14, 2022, www.thetrumparchive.com.

33 **In the Institute of Medicine's seminal:** Institute of Medicine (US) Committee on Quality of Health Care in America, *To Err Is Human: Building a Safer Health System*, ed. Linda T. Kohn, Janet M. Corrigan, and Molla S. Donaldson (Washington, DC: National Academies Press, 2000), http://www.ncbi.nlm.nih.gov/books/NBK225182.

36 **The use of ACE inhibitors:** Adriana Albini et al., "The SARS-CoV-2 Receptor, ACE-2, Is Expressed on Many Different Cell Types: Implications for ACE-Inhibitor- and Angiotensin II Receptor Blocker-Based Cardiovascular Therapies," *Internal and Emergency Medicine* 15, no. 5 (August 2020): 759–66, https://doi.org/10.1007/s11739-020-02364-6.

38 **It's big business:** Research and Markets, "Global Stem Cell Therapy Market Report 2021–2030: Allogeneic Stem Cell & Therapy Autologous Stem Cell Therapy/Adult Stem Cells, Induced Pluripotent Stem Cells, & Embryonic Stem Cells," news release, May 19, 2021, accessed February 16, 2022, https://www.prnewswire.com/news-releases/global-stem-cell-therapy-market-report-2021-2030-allogeneic-stem-cell--therapy-autologous-stem-cell-therapy--adult-stem-cells-induced-pluripotent-stem-cells--embryonic-stem-cells-301294924.html.

38 **In fact, a 2018 journal article by:** Michael J. Hayes et al., "Most Medical Practices Are Not Parachutes: A Citation Analysis of Practices Felt by Biomedical Authors to Be Analogous to Parachutes," *CMAJ Open* 6, no. 1 (March 2018): E31, https://doi.org/10.9778/cmajo.20170088.

40 **There is plenty of research showing:** Mahesh Chandra et al., "The Free Radical System in Ischemic Heart Disease," *International Journal of Cardiology* 43, no. 2 (February 1, 1994): 121–25, https://doi.org/10.1016/0167-5273(94)90001-9; Aaron J. Schetter, Niels H. H. Heegaard, and Curtis C. Harris, "Inflammation and Cancer: Interweaving MicroRNA, Free Radical, Cytokine and P53 Pathways," *Carcinogenesis* 31, no. 1 (January 1, 2010): 37–49, https://doi.org/10.1093/carcin/bgp272.

40 **Vitamin E pills didn't protect against:** HOPE and HOPE-TOO Trial Investigators, "Effects of Long-Term Vitamin E Supplementation on Cardiovascular Events and Cancer: A Randomized Controlled Trial," *JAMA* 293, no. 11 (March 1, 2005): 1338–47, https://doi.org/10.1001/jama.293.11.1338.

43 **In 1989, a study was published in:** Cardiac Arrhythmia Supression Trial (CAST) Investigators, "Preliminary Report: Effect of Encainide and Flecainide on Mortality in a Randomized Trial of Arrhythmia Suppression After Myocardial Infarction," *New England Journal of Medicine* 321, no. 6 (August 10, 1989): 406–12, https://doi.org/10.1056/NEJM198908103210629.

44 **When AZT, the first anti-HIV:** Margaret A. Fischl et al., "The Efficacy of Azidothymidine (AZT) in the Treatment of Patients with AIDS and AIDS-Related Complex," *New England Journal of Medicine* 317, no. 4 (July 23, 1987): 185–91, https://doi.org/10.1056/NEJM198707233170401.

CHAPTER 3: THE TEMPTATION OF THE "ONE SIMPLE THING"

52 **It's true that multiple studies:** Michael F. Scheier and Charles S. Carver, "Adapting to Cancer: The Importance of Hope and Purpose," in *Psychosocial Interventions for Cancer* (Washington, DC: American Psychological Association, 2001), 15–36, https://doi.org/10.1037/10402-002.

53 **"sense of impending doom":** Lynn Eldridge, "Is a Sense of Impending Doom a Real Medical Symptom?" Verywell Mind, November 5, 2021, accessed February 16, 2022, https://www.verywellmind.com/sense-of-impending-doom-symptom-4129656.

56 **The fact that an extract of kale:** Kosuke Nishi et al., "Immunostimulatory *in Vitro* and *in Vivo* Effects of a Water-Soluble Extract from Kale," *Bioscience, Biotechnology, and Biochemistry* 75, no. 1 (January 23, 2011): 40–46, https://doi.org/10.1271/bbb.100490.

57 **When compared to a more reliable:** E. A. Krall and J. T. Dwyer, "Validity of a Food Frequency Questionnaire and a Food Diary in a Short-Term Recall Situation," *Journal of the American Dietetic Association* 87, no. 10 (October 1, 1987): 1374–77.

58 **Take this example, which was presented:** Manpreet Kaur et al., "Impact of Chilli-Pepper Intake on All-Cause and Cardiovascular Mortality—a Systematic Review and Meta-Analysis," *Circulation* 142 (2020): A12729, https://doi.org/10.1161/circ.142.suppl_3.12729.

60 **In 2015, a study was published in:** Michael J. Orlich et al., "Vegetarian Dietary Patterns and the Risk of Colorectal Cancers," *JAMA Internal Medicine* 175, no. 5 (May 1, 2015): 767–76, https://doi.org/10.1001/jamainternmed.2015.59.

62 **In 2010, an ad for the cholesterol-lowering:** John Mack, "Guest Post: New Lipitor Ads Dis Exercise & Healthy Diet. Are You Kidding Me?" CardioBrief, November 7, 2010, http://www.cardiobrief.org/2010/11/07/guest-post-new-lipitor-ads-dis-exercise-healthy-diet-are-you-kidding-me.

65 **In 2017, researchers from Switzerland:** Katrin H. Preller et al., "The Fabric of Meaning and Subjective Effects in LSD-Induced States Depend on Serotonin 2A Receptor Activation," *Current Biology* 27, no. 3 (February 6, 2017): 451–57, https://doi.org/10.1016/j.cub.2016.12.030.

66 **Highly emotional, spiritual, religious:** Jacqueline Borg et al., "The Serotonin System and Spiritual Experiences," *American Journal of Psychiatry* 160, no. 11 (November 2003): 1965–69, https://doi.org/10.1176/appi.ajp.160.11.1965.

CHAPTER 4: THE QUEST FOR CAUSALITY

74 **Though it is unclear what led:** Michael E. Smith, *The Aztecs* (Malden, MA: Blackwell, 1997).

75 **It is quite possible that George:** Vibul V. Vadakan, "The Asphyxiating and Exsanguinating Death of President George Washington," *Permanente Journal* 8, no. 2 (Spring 2004): 76–79, https://www.thepermanentejournal.org/issues/2004/spring.html.

79 **In 1965, British epidemiologist Sir Austin Bradford Hill:** Austin Bradford Hill, "The Environment and Disease: Association or Causation?" *Proceedings of the Royal Society of Medicine* 58, no. 5 (1965), 295–300, https://doi.org/10.1177/003591576505800503.

80 **For example, in terms of the strength:** "15 Insane Things That Correlate with Each Other," Spurious Correlations, accessed February 16, 2022, http://tylervigen.com/spurious-correlations.

82 **Another example: In 1999, an article:** Graham E. Quinn et al., "Myopia and Ambient Lighting at Night," *Nature* 399 (May 1, 1999): 113–14, https://doi.org/10.1038/20094.

83 **In 2017, a study appeared in *Pediatrics*:** Lisa-Christine Girard, Orla Doyle, and Richard E. Tremblay, "Breastfeeding, Cognitive and Noncognitive Development in Early Childhood: A Population Study," *Pediatrics* 139, no. 4 (April 1, 2017): e20161848, https://doi.org/10.1542/peds.2016-1848.

84 **For example, it was a correlational:** L. L. Craven, "Acetylsalicylic Acid, Possible Preventive of Coronary Thrombosis," *Annals of Western Medicine and Surgery* 4, no. 2 (February 1950): 95.

85 **twenty-six thousand participants:** JoAnn E. Manson et al., "Vitamin D Supplements and Prevention of Cancer and Cardiovascular Disease," *New England Journal of Medicine* 380, no. 1 (January 3, 2019): 33–44, https://doi.org/10.1056/NEJMoa1809944.

85 **twenty-four hundred prediabetics:** Anastassios G. Pittas et al., "Vitamin D Supplementation and Prevention of Type 2 Diabetes," *New England Journal of Medicine* 381, no. 6 (August 8, 2019): 520–30.

85 **So a trial of 2,256 older women:** K. M. Sanders, A. L. Stuart, E. J. Williamson et al., "Annual High Dose Oral Vitamin D and Falls and Fractures in Older Women: A Randomized Controlled Trial," *JAMA* 303, no. 18 (2010): 1815–22, doi:10.1001/jama.2010.594.

85 **A Women's Health Initiative study randomized:** Andrea Z. LaCroix et al., "Calcium Plus Vitamin D Supplementation and Mortality in Postmenopausal Women: The Women's Health Initiative Calcium–Vitamin D

Randomized Controlled Trial," *Journals of Gerontology: Series A* 64A, no. 5 (May 1, 2009): 559–67, https://doi.org/10.1093/gerona/glp006.

88 **From a genetic perspective, humans:** Lynn B. Jorde, "Genetic Variation and Human Evolution," *American Society of Human Genetics* 7, no. 2019 (2003): 28–33.

88 **And, no, I don't only mean the overt:** Frank Edwards, Hedwig Lee, and Michael Esposito, "Risk of Being Killed by Police Use of Force in the United States by Age, Race–Ethnicity, and Sex," *Proceedings of the National Academy of Sciences* 116, no. 34 (August 20, 2019): 16793–98, https://doi.org/10.1073/pnas.1821204116.

91 **According to the 2013 National Survey:** National Survey on Drug Use and Health, accessed February 17, 2022, https://nsduhweb.rti.org/respweb/homepage.cfm.

92 **In 2009, researchers published:** Hans Olav Melberg, Andrew M. Jones, and Anne Line Bretteville-Jensen, "Is Cannabis a Gateway to Hard Drugs?" *Empirical Economics* 38, no. 3 (2010): 583–603. https://doi.org/10.1007/s00181-009-0280-z.

CHAPTER 5: HOW COIN FLIPS CHANGED MEDICINE FOREVER

101 **Just a few years earlier:** C. C. Lund and D. B. Mill, "Scurvy and Anson's Voyage Round the World, 1740–44: An Analysis of the Royal Navy's Worst Outbreak," *New England Journal of Medicine* (1940): 223, 363.

103 **Two received citrus fruit:** James Lind, *A Treatise on the Scurvy* (London: J. Millar, 1757).

105 **For example, in 2019, *JAMA Pediatrics*:** Rivka Green et al., "Association Between Maternal Fluoride Exposure During Pregnancy and IQ Scores in Offspring in Canada," *JAMA Pediatrics* 173, no. 10 (October 1, 2019): 940–48, https://doi.org/10.1001/jamapediatrics.2019.1729.

114 **For example, in 2014, the results of a randomized trial:** Abhinav Goyal et al., "Melatonin Supplementation to Treat the Metabolic Syndrome: A Randomized Controlled Trial," *Diabetology & Metabolic Syndrome* 6, no. 1 (November 18, 2014): 124, https://doi.org/10.1186/1758-5996-6-124.

117 **The starkest example I've seen of this was:** Frances Campbell et al., "Early Childhood Investments Substantially Boost Adult Health," *Science* 343, no. 6178 (March 28, 2014): 1478–85, https://doi.org/10.1126/science.1248429.

119 **With rare exceptions:** Klaus Linde et al., "How Large Are the Nonspecific Effects of Acupuncture? A Meta-Analysis of Randomized Controlled Trials," *BMC Medicine* 8, no. 1 (November 23, 2010): 75, https://doi.org/10.1186/1741-7015-8-75.

CHAPTER 6: GOOD MEDICINE MAY NOT BE GOOD FOR YOU

124 **The vast majority of mushroom toxicity:** B. Zane Horowitz and Michael J. Moss, "Amatoxin Mushroom Toxicity," in *StatPearls* (Treasure Island, FL: StatPearls Publishing, 2022), http://www.ncbi.nlm.nih.gov/books/NBK 431052.

132 **SPRINT had enrolled almost ten thousand people:** "A Randomized Trial of Intensive versus Standard Blood-Pressure Control," *New England Journal of Medicine* 373, no. 22 (November 26, 2015): 2103–16, https://doi.org/10.1056 /NEJMoa1511939.

134 **We may know quite precisely that if we treat:** "Indications for Fibrinolytic Therapy in Suspected Acute Myocardial Infarction: Collaborative Overview of Early Mortality and Major Morbidity Results from All Randomised Trials of More Than 1000 Patients. Fibrinolytic Therapy Trialists' (FTT) Collaborative Group," *Lancet* 343, no. 8893 (February 5, 1994): 311–22.

136 **A study that appeared in *JAMA Network Open*:** Daniel J. Morgan et al., "Clinician Conceptualization of the Benefits of Treatments for Individual Patients," *JAMA Network Open* 4, no. 7 (July 21, 2021): e2119747, https://doi .org/10.1001/jamanetworkopen.2021.19747.

CHAPTER 7: NO SUCH THING AS INCURABLE

149 **In fact, the desperation for an effective:** Lisa Urquhart, "The Industry's Biggest Drug Launches," *Nature Reviews Drug Discovery* 17, no. 12 (2018): 855, doi:10.1038/nrd.2018.209.

150 **The first treatment for hepatitis C:** Lorena Puchades Renau and Marina Berenguer, "Introduction to Hepatitis C Virus Infection: Overview and History of Hepatitis C Virus Therapies," *Hemodialysis International* 22, no. S1 (2018): S8–S21, doi:10.1111/hdi.12647.

152 **In fact, in October 2020:** F. Perry Wilson, "What If the Covid-19 Vaccine Only Works Half the Time?" *Vox*, October 22, 2020.

154 **For example, a study published in September 2020 found:** Paul Bertin, Kenzo Nera, and Sylvain Delouvée, "Conspiracy Beliefs, Rejection of Vaccination, and Support for Hydroxychloroquine: A Conceptual Replication-Extension in the COVID-19 Pandemic Context," *Frontiers in Psychology* 11 (2020), https://www.frontiersin.org/article/10.3389/fpsyg.2020.565128.

154 **But a paper published in the journal *Medical Hypotheses*:** Robert F. Cathcart III, "Vitamin C in the Treatment of Acquired Immune Deficiency Syndrome (AIDS)," *Medical Hypotheses* 14, no. 4 (August 1, 1984): 423–33, https:// doi.org/10.1016/0306-9877(84)90149-X.

156 **Even so, only about five out of one hundred medicines:** Helen Dowden and Jamie Munro, "Trends in Clinical Success Rates and Therapeutic Focus," *Nature Reviews Drug Discovery* 18, no. 7 (May 8, 2019): 495–96, https://doi.org/10.1038/d41573-019-00074-z.

158 **Early in the pandemic, some researchers:** Kristina M. Cross et al., "Melatonin for the Early Treatment of COVID-19: A Narrative Review of Current Evidence and Possible Efficacy," *Endocrine Practice* 27, no. 8 (August 2021): 850, https://doi.org/10.1016/j.eprac.2021.06.001.

CHAPTER 8: PHARMA

164 **Starting around 2016, study after study:** Giovanni Antonio Silverii, Matteo Monami, and Edoardo Mannucci, "Sodium-Glucose Co-Transporter-2 Inhibitors and All-Cause Mortality: A Meta-Analysis of Randomized Controlled Trials," *Diabetes, Obesity and Metabolism* 23, no. 4 (2021): 1052–56, https://doi.org/10.1111/dom.14286.

167 **In 2019, a Gallup poll found:** Justin McCarthy, "Big Pharma Sinks to the Bottom of U.S. Industry Rankings," Gallup, September 3, 2019, https://news.gallup.com/poll/266060/big-pharma-sinks-bottom-industry-rankings.aspx.

168 **In 2021, the pharmaceutical industry spent:** "Industries," OpenSecrets, accessed February 17, 2022, https://www.opensecrets.org/federal-lobbying/industries?cycle=2021.

169 **When criticized for the high prices of approved:** Rebecca Farley, "Do Pharmaceutical Companies Spend More on Marketing Than Research and Development?" *PharmacyChecker*, April 28, 2021, https://www.pharmacychecker.com/askpc/pharma-marketing-research-development/.

169 **A 2019 poll found that 29 percent of adults:** Ashley Kirzinger et al., "KFF Health Tracking Poll—February 2019: Prescription Drugs," Kaiser Family Foundation, March 1, 2019, https://www.kff.org/health-costs/poll-finding/kff-health-tracking-poll-february-2019-prescription-drugs.

169 **According to the West Health Policy Center:** West Health, "New Study Predicts More Than 1.1 Million Deaths Among Medicare Recipients Due to the Inability to Afford Their Medications," news release, November 19, 2020, accessed February 17, 2022, https://www.westhealth.org/press-release/study-predicts-1-million-deaths-due-to-high-cost-prescription-drugs.

172 **The who-knew-what-when history**: Jerry Avorn, MD, *Powerful Medicines: The Benefits, Risks, and Costs of Prescription Drugs* (New York: Doubleday, 2008).

170 **A study published in *PLOS Medicine* found:** G. Michael Allan, Joel Lexchin, and Natasha Wiebe, "Physician Awareness of Drug Cost: A Systematic

Review," *PLOS Medicine* 4, no. 9 (September 2007), https://doi.org/10.1371/journal.pmed.0040283.

173 **The risk of heart attack was four times higher:** "PDF FileCase 2:05-Cv-02367-SRC-CLW. UNITED STATES DISTRICT COURT DISTRICT OF NEW JERSEY IN RE MERCK & CO," vdocuments.net, accessed July 26, 2022, https://vdocuments.net/in-re-merck-co-inc-securities-derivative-erisa-case-205-cv-02367-src-clw.html.

174 **Writing in the *Lancet* in 2005:** David J. Graham et al., "Risk of Acute Myocardial Infarction and Sudden Cardiac Death in Patients Treated with Cyclo-Oxygenase 2 Selective and Non-Selective Non-Steroidal Anti-Inflammatory Drugs: Nested Case-Control Study," *Lancet* 365, no. 9458 (February 5, 2005): 475–81, https://doi.org/10.1016/S0140-6736(05)17864-7.

175 **At that time, you could buy:** Ed Silverman, "Mylan CEO Accepts Full Responsibility for EpiPen Price Hikes, but Offers Little Explanation," *STAT*, December 1, 2016, https://www.statnews.com/pharmalot/2016/12/01/mylan-ceo-responsibility-epipen-price.

175 **Mylan's CEO, testifying to Congress:** Ibid.

176 **It's true that pharma spends an inordinate:** Kathryn R. Tringale et al., "Types and Distribution of Payments from Industry to Physicians in 2015," *JAMA* 317, no. 17 (May 2, 2017): 1774–84, https://doi.org/10.1001/jama.2017.3091.

176 **To be fair, some physicians are taking:** Charles Ornstein, Tracy Weber, and Ryann Grochowski Jones, "We Found Over 700 Doctors Who Were Paid More Than a Million Dollars by Drug and Medical Device Companies," ProPublica, October 17, 2019, accessed February 17, 2022, https://www.propublica.org/article/we-found-over-700-doctors-who-were-paid-more-than-a-million-dollars-by-drug-and-medical-device-companies.

176 **In 2021, researchers published a paper in:** Aaron P. Mitchell et al., "Are Financial Payments from the Pharmaceutical Industry Associated with Physician Prescribing?" *Annals of Internal Medicine* 174, no. 3 (March 16, 2021): 353–61, https://doi.org/10.7326/M20-5665.

177 **In fact, a 2019 study in *JAMA Network Open*:** Genevieve P. Kanter et al., "US Nationwide Disclosure of Industry Payments and Public Trust in Physicians," *JAMA Network Open* 2, no. 4 (April 12, 2019): e191947, https://doi.org/10.1001/jamanetworkopen.2019.1947.

178 **They have continued more or less unabated:** D. C. Marshall, E. S. Tarras, K. Rosenzweig, D. Korenstein, and S. Chimonas, "Trends in Industry Payments to Physicians in the United States from 2014 to 2018," *JAMA* 324, no. 17 (2020): 1785–88, doi:10.1001/jama.2020.11413.

180 **Humira costs about $3,000 per:** Zachary Brennan, "House Committee Uncovers How Humira's Price Spiked by 470% as AbbVie Execs Cashed Bonuses Tied to the Hikes," *Endpoints News*, May 18, 2021, accessed February 17, 2022, https://endpts.com/house-committee-uncovers-how-humiras-price-spiked-by-470-as-abbvie-execs-cashed-bonuses-tied-to-the-hikes.

181 **In 2019, an FDA report documented that:** Center for Drug Evaluation and Research, "Generic Competition and Drug Prices," US Food & Drug Administration, December 13, 2019, https://www.fda.gov/about-fda/center-drug-evaluation-and-research-cder/generic-competition-and-drug-prices.

181 **But according to a 2017 report:** "Competition in Generic Drug Markets," National Bureau of Economic Research, November 2017, accessed February 17, 2022, https://www.nber.org/digest/nov17/competition-generic-drug-markets.

181 **As of December 2021, these include nitroglycerin:** "List of Off-Patent, Off-Exclusivity Drugs Without an Approved Generic," US Food & Drug Administration, December 16, 2021, accessed June 13, 2022, https://www.fda.gov/drugs/abbreviated-new-drug-application-anda/list-patent-exclusivity-drugs-without-approved-generic.

183 **In a 2021 Kaiser Family Foundation poll, 88 percent:** Ashley Kirzinger et al., "KFF Health Tracking Poll—May 2021: Prescription Drug Prices Top Public's Health Care Priorities," Kaiser Family Foundation, June 3, 2021, https://www.kff.org/health-costs/poll-finding/kff-health-tracking-poll-may-2021.

184 **In 2007, well-known medical researcher Steven Schroeder:** Steven A. Schroeder, "We Can Do Better—Improving the Health of the American People," *New England Journal of Medicine* 357, no. 12 (September 20, 2007): 1221–28, https://doi.org/10.1056/NEJMsa073350.

CHAPTER 9: TOO GOOD TO BE TRUE

186 **In 2001, an article appeared in the science:** Eric Nagourney, "Vital Signs: Fertility; a Study Links Prayer and Pregnancy," *New York Times*, October 2, 2001, https://www.nytimes.com/2001/10/02/health/vital-signs-fertility-a-study-links-prayer-and-pregnancy.html.

186 **The study described a group of around:** Kwang Y. Cha, Daniel P. Wirth, and Rogerio A. Lobo, "Does Prayer Influence the Success of in Vitro Fertilization–Embryo Transfer? Report of a Masked, Randomized Trial," *Journal of Reproductive Medicine* 46, no. 9 (September 2001): 781–87.

190 **In the National Academy of Sciences work:** University of Chicago faculty committee, "Procedures for Investigating Academic Fraud," in *Responsible*

Science: Ensuring the Integrity of the Research Process: Volume II, ed. National Academy of Sciences, National Academy of Engineering, and Institute of Medicine (Washington, DC: National Academies Press, 1993), 250. Available from https://www.ncbi.nlm.nih.gov/books/NBK236197/.

192 **Governmental agencies have conducted:** Stephen L. George and Marc Buyse, "Data Fraud in Clinical Trials," *Clinical Investigation* 5, no. 2 (2015): 161–73, https://doi.org/10.4155/cli.14.116.

192 **A survey of a diverse group of research:** Erica R. Pryor, Barbara Habermann, and Marion E. Broome, "Scientific Misconduct from the Perspective of Research Coordinators: A National Survey," *Journal of Medical Ethics* 33, no. 6 (June 1, 2007): 365–69, https://doi.org/10.1136/jme.2006.016394.

192 **At the high end of the estimates:** Stephen L. George and Marc Buyse, "Data Fraud in Clinical Trials," *Clinical Investigation* 5, no. 2 (2015): 161–73, https://doi.org/10.4155/cli.14.116.

194 **Penalties for research misconduct can:** Matt Miller, "Ex-PSU Professor Craig Grimes Sentenced to Federal Prison for Research Grant Fraud," PennLive, November 30, 2012, https://www.pennlive.com/midstate/2012/11/ex-psu_prof_craig_grimes_sente.html.

195 **The study was titled:** A. J. Wakefield et al., "RETRACTED: Ileal-Lymphoid-Nodular Hyperplasia, Non-Specific Colitis, and Pervasive Developmental Disorder in Children," *Lancet* 351, no. 9103 (February 28, 1998): 637–41, https://doi.org/10.1016/S0140-6736(97)11096-0.

197 **In a series of articles:** Brian Deer, "How the Vaccine Crisis Was Meant to Make Money," *BMJ* 342 (January 11, 2011): c5258, https://doi.org/10.1136/bmj.c5258.

197 **According to a private prospectus leaked:** Ibid.

197 **Deer subsequently identified the twelve:** Ibid.

197 **It took more than two decades:** Jonathan D. Quick and Heidi Larson, "The Vaccine-Autism Myth Started 20 Years Ago. Here's Why It Still Endures Today," *Time*, February 28, 2018, accessed February 17, 2022, https://time.com/5175704/andrew-wakefield-vaccine-autism.

202 **In 2005, the *Lancet* published:** J. Sudbø et al., "RETRACTED: Non-Steroidal Anti-Inflammatory Drugs and the Risk of Oral Cancer: A Nested Case-Control Study," *Lancet* 366, no. 9494 (October 15, 2005): 1359–66, https://doi.org/10.1016/S0140-6736(05)67488-0.

202 **All 908 subjects in the study were fictitious:** Stephen L. George and Marc Buyse, "Data Fraud in Clinical Trials," *Clinical Investigation* 5, no. 2 (2015): 161–73, https://doi.org/10.4155/cli.14.116.

204 **But one study stood out for the profound effect size:** Ahmed Elgazzar et al., "Efficacy and Safety of Ivermectin for Treatment and Prophylaxis of COVID-19 Pandemic" (2020), https://doi.org/10.21203/rs.3.rs-100956/v3.

206 **In that case, the investigators:** Xavier Nogués et al., "Calcifediol Treatment and COVID-19-Related Outcomes," *SSRN Scholarly Paper* (Rochester, NY: Social Science Research Network, January 22, 2021), https://doi.org/10.2139/ssrn.3771318.

207 **What they found was:** Melissa Davey, "Huge Study Supporting Ivermectin as Covid Treatment Withdrawn over Ethical Concerns," *Guardian*, July 15, 2021, https://www.theguardian.com/science/2021/jul/16/huge-study-supporting-ivermectin-as-covid-treatment-withdrawn-over-ethical-concerns.

207 **I understand why a huge pharmaceutical company:** Snigdha Prakash and Vikki Valentine, "Timeline: The Rise and Fall of Vioxx," NPR, November 10, 2007, https://www.npr.org/2007/11/10/5470430/timeline-the-rise-and-fall-of-vioxx.

210 **Lives would have been saved:** Gilmar Reis et al., for the TOGETHER Investigators, "Effect of Early Treatment with Ivermectin Among Patients with COVID-19," *New England Journal of Medicine* 386 (May 5, 2022), 1721–31, http://doi.org/10.1056/NEJMoa2115869.

CHAPTER 10: HEALING THE SYSTEM

218 **As of 2021, 70 percent of all physicians:** Laura Dyrda, "70% of Physicians Are Now Employed by Hospitals or Corporations," *Becker's ASC Review*, July 1, 2021, accessed February 17, 2022, https://www.beckersasc.com/asc-transactions-and-valuation-issues/70-of-physicians-are-now-employed-by-hospitals-or-corporations.html.

218 **In 2019, *Healthline* crunched the:** Heather Ross, "The Great Healthcare Bloat: 10 Administrators for Every 1 U.S. Doctor," *Healthline*, January 30, 2019, https://www.healthline.com/health-news/policy-ten-administrators-for-every-one-us-doctor-092813.

218 **The *Harvard Business Review* looked at the:** Robert Kocher, "The Downside of Health Care Job Growth," *Harvard Business Review*, September 23, 2013, https://hbr.org/2013/09/the-downside-of-health-care-job-growth.

220 **just 8.6 percent:** "Physician Pay Accounts for 8.6% of Total Healthcare Expenses," *Becker's Hospital Review*, May 16, 2012, accessed February 17, 2022, https://www.beckershospitalreview.com/compensation-issues/physician-pay-accounts-for-86-of-total-healthcare-expenses.html.

220 **A recent social media trend**: "r/HospitalBills," Reddit, accessed February 17, 2022, https://www.reddit.com/r/HospitalBills.

222 **Data suggests that the system has lowered costs:** "Hospital Rate Setting: Successful in Maryland but Challenging to Replicate," Altarum Healthcare Value Hub, May 2020, accessed February 17, 2022, https://www.healthcarevalue

hub.org/advocate-resources/publications/hospital-rate-setting-promising-challenging-replicate.

222 **Maryland, the seventh-most-expensive state:** "Average Cost of Health Insurance (2022)," ValuePenguin, accessed June 10, 2022, https://www.valuepenguin.com/average-cost-of-health-insurance.

224 **The term "VIP syndrome" was:** Walter Weintraub, " 'The VIP Syndrome': A Clinical Study in Hospital Psychiatry," *Journal of Nervous and Mental Disease* 138, no. 2 (February 1964): 181–93.

225 **In 2019, researchers from Washington University in St. Louis:** Alan Ge and David L. Brown, "Assessment of Cardiovascular Diagnostic Tests and Procedures Offered in Executive Screening Programs at Top-Ranked Cardiology Hospitals," *JAMA Internal Medicine* 180, no. 4 (April 1, 2020): 586–89, https://doi.org/10.1001/jamainternmed.2019.6607.

226 **In a 2014 piece in *JAMA*:** Michael B. Rothberg, "The $50,000 Physical," *JAMA* 311, no. 21 (June 4, 2014): 2175–76, https://doi.org/10.1001/jama.2014.3415.

228 **It was in that year that doctors Barry Marshall:** Barry J. Marshall and J. Robin Warren, "Unidentified Curved Bacilli in the Stomach of Patients with Gastritis and Peptic Ulceration," *Lancet* 1, no. 8390 (June 16, 1984): 1311–15, https://doi.org/10.1016/s0140-6736(84)91816-6.

229 **The year 2021 saw a treasure trove:** Tara Subramaniam, "The Big Takeaways from the Facebook Papers," CNN Business, October 26, 2021, accessed February 17, 2022, https://www.cnn.com/2021/10/26/tech/facebook-papers-takeaways/index.html.

229 **These emojis were weighted five times:** "Facebook Prioritized 'Angry' Emoji Reaction Posts in News Feeds," *Washington Post*, October 26, 2021, https://www.washingtonpost.com/technology/2021/10/26/facebook-angry-emoji-algorithm/.

230 **A 2010 article in the journal *Health Affairs*:** Michael V. Maciosek et al., "Greater Use of Preventive Services in U.S. Health Care Could Save Lives at Little or No Cost," *Health Affairs* 29, no. 9 (September 2010): 1656–60, https://doi.org/10.1377/hlthaff.2008.0701.

CHAPTER 11: MOVING TOGETHER

240 **Beginning around 2000, epidemiologists:** Anne Case and Angus Deaton, "Mortality and Morbidity in the 21st Century," *Brookings Papers on Economic Activity* (Spring 2017): 397–476, https://www.brookings.edu/bpea-articles/mortality-and-morbidity-in-the-21st-century.

240 **While drug overdoses in the setting of the opioid epidemic:** United States Congress, Joint Economic Committee Republicans, "Long-Term Trends in Deaths of Despair," September 5, 2019, accessed June 16, 2022,

https://www.jec.senate.gov/public/index.cfm/republicans/2019/9/long-term-trends-in-deaths-of-despair.

241 **Loneliness has a powerful effect:** John T. Cacioppo and Stephanie Cacioppo, "Social Relationships and Health: The Toxic Effects of Perceived Social Isolation," *Social and Personality Psychology Compass* 8, no. 2 (February 2014): 58–72, https://doi.org/10.1111/spc3.12087.

241 **A large meta-analysis, combining:** Julianne Holt-Lunstad et al., "Loneliness and Social Isolation as Risk Factors for Mortality: A Meta-Analytic Review," *Perspectives on Psychological Science* 10, no. 2 (March 1, 2015): 227–37, https://doi.org/10.1177/1745691614568352.

241 **What's worse, the healthy way to cope:** "How Do Americans Cope with Loneliness?" Kaiser Family Foundation, November 21, 2018, https://www.kff.org/other/slide/how-do-americans-cope-with-loneliness.

250 **It could be, though; there is:** Aaron Lazare, "Apology in Medical Practice: An Emerging Clinical Skill," *JAMA* 296, no. 11 (September 20, 2006): 1401–04, https://doi.org/10.1001/jama.296.11.1401.

252 **These laws are rather weak and probably:** Benjamin J. McMichael, R. Lawrence Van Horn, and W. Kip Viscusi, " 'Sorry' Is Never Enough: How State Apology Laws Fail to Reduce Medical Malpractice Liability Risk," *Stanford Law Review* 71, no. 2 (February 2019): 341–409, https://papers.ssrn.com/sol3/papers.cfm?abstract_id=2883693.

About the Author

F. PERRY WILSON grew up in Connecticut before attending Harvard University, where he graduated with honors in biochemistry. He then attended medical school at Columbia University College of Physicians and Surgeons before completing his internship, residency, and fellowship at the University of Pennsylvania. In 2012, he received a master's degree in clinical epidemiology, which has informed his research ever since. At Yale University since 2014, his goal is to use patient-level data and advanced analytics to personalize medicine to each individual patient. To that end, he holds multiple grants from the National Institutes of Health and is the director of Yale's Clinical and Translational Research Accelerator.

Dr. Wilson is internationally recognized for his expertise in the design and interpretation of medical studies. He has appeared on CNN, HLN, and NPR and has written for MedPage Today, Medscape, *HuffPost*, and the *Milwaukee Journal Sentinel*, among others. His free online course, entitled Understanding Medical Research: Your Facebook Friend Is Wrong, is one of the most popular courses on Coursera.